T0202340

The World in the Wave Function

The World in the Wave Function

A Metaphysics for Quantum Physics

ALYSSA NEY

OXFORD
UNIVERSITY PRESS

Oxford University Press is a department of the University of Oxford. It furthers
the University's objective of excellence in research, scholarship, and education
by publishing worldwide. Oxford is a registered trade mark of Oxford University
Press in the UK and certain other countries.

Published in the United States of America by Oxford University Press
198 Madison Avenue, New York, NY 10016, United States of America.

Library of Congress Cataloging-in-Publication Data
Names: Ney, Alyssa, author.
Title: The world in the wave function : a metaphysics
for quantum physics / Alyssa Ney.
Description: New York, NY : Oxford University Press, [2021] |
Includes bibliographical references and index.
Identifiers: LCCN 2020024392 (print) | LCCN 2020024393 (ebook) |
ISBN 9780190097714 (hardback) | ISBN 9780190097738 (epub)
Subjects: LCSH: Wave functions. | Quantum theory.
Classification: LCC QC174.26.W3 N49 2020 (print) | LCC QC174.26.W3 (ebook)
| DDC 530.12/4—dc23
LC record available at https://lccn.loc.gov/2020024392
LC ebook record available at https://lccn.loc.gov/2020024393

DOI: 10.1093/oso/9780190097714.001.0001

1 3 5 7 9 8 6 4 2

Printed by Integrated Books International, United States of America

On this day, however, the listener saw something odd when he glanced at the waveform display. Even experts had a hard time telling with the naked eye whether a waveform carried information. But the listener was so familiar with the noise of the universe that he could tell that the wave that now moved in front of his eyes had something extra. The thin curve, rising and falling, seemed to possess a soul.

—Liu Cixin, *The Three-Body Problem* (2006)

Contents

Preface   ix
Acknowledgments   xiii

1. A Preliminary Case for Wave Function Realism   1
   1.1   Wave Function Representations in Quantum
         Mechanics   1
   1.2   The Measurement Problem   14
   1.3   Orthodox Quantum Mechanics   15
   1.4   Quantum Mechanics without Measurement   25
   1.5   Wave Function Realism   33
   1.6   A Higher-Dimensional Reality   47

2. The Argument from Entanglement   49
   2.1   Entanglement as the Characteristic Feature of
         Quantum Theories   49
   2.2   The Necessity of Wave Function Realism?   52
   2.3   Rivals: The Primitive Ontology Approach   56
   2.4   Rivals: Holisms   63
   2.5   Rivals: Relational Approaches   67
   2.6   Rivals: Spacetime State Realism   72
   2.7   Rivals: The Multi-Field Approach   74
   2.8   The Contingency of Wave Function Realism   76

3. The Virtues of Separability and Locality   80
   3.1   The Case for Wave Function Realism   80
   3.2   Separability   81
   3.3   Separability and Wave Function Realism   87
   3.4   A Challenge   90
   3.5   Concepts of Locality   96
   3.6   Quantum Nonlocality   98
   3.7   Locality and Wave Function Realism   104
   3.8   Avoiding Nonlocality with Nonseparability   113
   3.9   Motivating a Separable and Local Metaphysics   120
   3.10 In Defense of Intuitions   129

4. Wave Function Realism in a Relativistic Setting 133
   4.1 Removing Idealization 133
   4.2 Five Critiques 135
   4.3 Wave Function Realism for Relativistic Quantum
       Theories 138
   4.4 Interpretations and Interpretational Frameworks 149
   4.5 Response to Objections 150
   4.6 Wave Function Realism in the Limit of Physical
       Theorizing 160

5. Must an Ontology for Quantum Theories Contain
   Local Beables? 166
   5.1 The Constitution Objection 166
   5.2 Doing without Macroscopic Objects 169
   5.3 The Threat of Empirical Incoherence 173
   5.4 Primitive Ontologies and Local Beables 181
   5.5 Perception and the Macroscopic 194

6. The Causal Role of Macroscopic Objects 197
   6.1 The Macro-Object Problem 197
   6.2 An Initial Proposal 199
   6.3 Monton's Challenge 207
   6.4 Albert's Proposal 210
   6.5 Troubles with Functionalism 216
   6.6 The Decoherence Strategy 219
   6.7 From Simulation to Constitution 223

7. Finding the Macroworld 225
   7.1 A Constitutive Explanation in Two Stages 225
   7.2 The Role of Grounding 226
   7.3 Recovering Three-Dimensionality Using Symmetries 231
   7.4 Parts and Wholes 238
   7.5 Partial Instantiation 241
   7.6 Tables, Chairs, and the Rest 247
   7.7 Finding the World in the Wave Function 248

Postscript: An Incredulous Stare 251
References 255
Index 265

Preface

If quantum theories are true, what kind of truth do they suggest? What are their ontological implications, that is, what do they tell us about the fundamental objects that make up our world? How should quantum theories make us reevaluate our classical conceptions of the basic constitution of material objects and ourselves? And what lessons do they carry for the way objects may interact with one another? A century after the development of quantum theories, there is still nothing like a consensus answer to any of these questions, nor even a received view. And yet it is natural to wonder what these theories that have been so remarkably empirically successful may be telling us about ourselves and the world that surrounds us.

The goal of this book is to develop and defend one framework for understanding the kind of world described by quantum theories. This is a framework initially suggested by the wave representation for quantum theories developed by Erwin Schrödinger in the 1920s, but only much later explicitly proposed and defended as an account of reality in the work of the philosophers of physics David Albert and Barry Loewer in the 1990s. Albert characterized it as the necessary point of view for those purporting to be realists about quantum theories (1996, p. 277), that is, for those who regard quantum theories as approximately correct and objective representations of our world, rather than merely useful mathematical tools to predict the results of future experiments. This framework is what Albert and Loewer have called *wave function realism*. It is a way of interpreting quantum theories so that the central object they describe is the quantum wave function, an object they view as a field on an extremely high-dimensional space.

According to wave function realism, we and all of the objects around us are ultimately constituted out of the wave function and although we may seem to occupy a three-dimensional space of the kind described by classical physics, the more fundamental spatial framework of quantum worlds like ours is instead quite different, one of very many dimensions, with no three of these dimensions corresponding to the heights, widths, and depths of our ordinary experience.

In this book, I will not try to make the case that wave function realism is a *necessary* framework one must adopt if one is to be a realist about quantum theories.[1] In the past few decades, several realist frameworks have been proposed, each presenting us with an interesting way of understanding the world described by quantum theories that is worthy of development. As we will see, each of these alternative frameworks has its own advocates who have made compelling cases in defense of their preferred interpretations. Actually, because the more common attitude about quantum theories for a very long time was that one could not or should not understand them realistically, and because such attitudes are still very much entrenched in large pockets of the physics community, we are still at an early stage of the conversation about what might be the correct understanding of the ontological implications of quantum theories. For this reason, there is still much work to be done, in spelling out the details of these frameworks, and what may be said in their defense. And so, while my main task here will be to make it clear that wave function realism is worth taking seriously as a framework for

---

[1] Although the debate between realists and anti-realists will sometimes come up in the chapters that follow, my main task in this book will not be to argue for realism about quantum theories. Instead, my approach will be to take realism as a starting point and see what can be said for this one way of developing a realist approach, wave function realism. The work of defending realism about quantum theories has already been satisfactorily addressed by others, in my view. If one is not yet convinced, I especially recommend the first chapter of David Wallace's 2012 *The Emergent Multiverse* as well as the second chapter of Peter Lewis's 2016 *Quantum Ontology*. Adam Becker's 2018 *What Is Real?* provides useful historical background for the debates about realism in quantum mechanics.

understanding the worlds described by our best quantum theories, my stance in this book will be one of humility and tolerance for other approaches.[2]

While wave function realism has been defended for more than twenty years now, it is still a rather minority position among those working in physics and philosophy of physics.[3] Even among those who would be realists about quantum theories, the view has met a variety of criticisms that have struck many working in the field as decisive. Wave function realism has been variously criticized as philosophically naïve and unmotivated, as applicable only as an understanding of the most basic and idealized quantum systems, and as empirically incoherent for its failure to allow a plausible account of how low-dimensional material objects, of the kind we perceive around us and in which the experimental results confirming quantum theories have been recorded, may ultimately be constituted out of a wave function inhabiting a high-dimensional space. And so, wave function realism is a position whose advocates

---

[2] Here I follow Rudolf Carnap:

> The acceptance or rejection of abstract linguistic forms, just as the acceptance or rejection of any other linguistic forms in any branch of science, will finally be decided by their efficiency as instruments, the ratio of the results achieved to the amount and complexity of the efforts required. To decree dogmatic prohibitions of certain linguistic forms instead of testing them by their success or failure in practical use, is worse than futile; it is positively harmful because it may obstruct scientific progress. . . . Let us grant to those who work in any special field of investigation the freedom to use any form of expression which seems useful to them; the work in the field will sooner or later lead to the elimination of those forms which have no useful function. *Let us be cautious in making assertions and critical in examining them, but tolerant in permitting linguistic forms.* (1947, p. 221)

as well as some of my main interlocutors:

> While we argue for the adoption of spacetime state realism over wave-function realism, we wish to remain neutral on whether one of these (or perhaps some third) really does provide the One True Interpretation of the quantum state, or whether one is merely a more perspicuous description than the other, a description of something that we are ultimately unable to render unequivocally in intuitive terms. (Wallace and Timpson 2010, p. 701)

[3] Although perhaps due to the influence of some its main proponents, it has had something of an outsized acceptance in metaphysical circles.

within physics and the philosophy of physics can be counted on a single hand. It is my contention that this is because the best case for wave function realism has until now not been clearly stated, nor has it been shown how the framework is to be applied to anything other than the formalism of the most simple nonrelativistic quantum mechanics. In the pages that follow, I present what I regard as the best case that can be made for wave function realism. I also develop the framework so that it may be applied to more sophisticated relativistic quantum theories, including quantum field theories. In the final chapters, I develop an account of how we may ultimately see ourselves and other material objects as constituted out of wave function stuff.

Throughout the book, I make an effort to avoid mathematical and technical jargon beyond basic calculus where possible. Where it is necessary, such jargon is explained. Although I expect the main audience for this book will comprise physicists and philosophers working on the interpretation of quantum theories, I intend this book to be accessible to the intelligent layperson. For, as we will see, in addition to being surprising and fascinating, one of the virtues of wave function realism is that it is a framework for understanding quantum theories that, while reflecting the inescapable weirdness of quantum theories, is particularly intelligible. Indeed, this is one of the chief lessons that Albert and Loewer wished to put forward in presenting wave function realism in the first place, that early interpreters were simply wrong to think one couldn't gain a clear understanding of the world or worlds presented by quantum theories. As we will see, there are many ways to understand the worlds our best quantum theories describe, and each competitor framework for interpretation accommodates quantum weirdness in its own way, including wave function realism with its higher dimensions. To say something is weird is not to say that it cannot be understood. Wave function realism affords one accessible, intelligible, even visualizable way of doing so.

Acknowledgments

This work was made possible due to the support of the National Science Foundation under Grant No. 1632546. I thank the University of California, Davis for allowing me the sabbatical years 2016–17 and 2018–19 to complete this project under that grant and an Academic Cross-Training Fellowship from the John Templeton Foundation. I also thank the University of Rochester for a research leave that allowed me to visit Columbia University in the Fall of 2008. I owe immense gratitude to my two dear mentors, David Albert and Barry Loewer, for many discussions and for the inspiration to take up this project. Thanks also to Valia Allori, Peter Lewis, Wayne Myrvold, Jill North, Chip Sebens, Paul Teller, and David Wallace for reading my work and providing invaluable feedback over the years. I would also like to thank my editor, Peter Ohlin, for his support and guidance.

Some of the arguments below have appeared in previously published work:

Ney, Alyssa and Kathryn Phillips. 2013. Does an Adequate Physical Theory Demand a Primitive Ontology? *Philosophy of Science*. 80: 454–474.

Ney, Alyssa. 2015. Fundamental Physical Ontologies and the Constraint of Empirical Coherence: A Defense of Wave Function Realism. *Synthese*. 192(10): 3105–3124.

Ney, Alyssa. 2017 (Online first). Finding the World in the Wave Function: Some Strategies for Solving the Macro-object Problem. *Synthese*. doi:10.1007/s11229-017-1349-4.

Ney, Alyssa. 2019. Locality and Wave Function Realism. *Quantum Worlds: Perspectives on the Ontology of Quantum Mechanics*. O.

Lombardi, S. Fortin, C. López, and F. Holik, eds. Cambridge: Cambridge University Press, 164–182.

Ney, Alyssa. 2020. Wave Function Realism in a Relativistic Setting. *The Foundation of Reality*. G. Darby, D. Glick, and A. Marmodoro, eds. Oxford: Oxford University Press, 154–168.

Ney, Alyssa. 2020. Separability, Locality, and Higher Dimensions in Quantum Mechanics. *Current Controversies in Philosophy of Science*. S. Dasgupta and B. Weslake, eds. London: Routledge, 75–90.

The World in the Wave Function

1

A Preliminary Case for Wave Function Realism

1.1 Wave Function Representations in Quantum Mechanics

The central question to which this book is addressed is the question of what may be seen as the ontological implications of quantum theories. Questions about ontology are questions about the kinds of things that exist according to one theory or other. And so, here we will ask: what is the character of our world if quantum theories are true, what fundamentally exists according to these theories, and what sorts of things do these fundamental objects constitute?

Sometimes these questions get conflated with another question that has been the subject of far more debate in the foundations of physics. This is the question of the correct solution to the measurement problem. Both the ontological questions and the measurement problem concern the proper interpretation of quantum theories. But the measurement problem is not primarily a question about ontology. Rather it concerns what may be seen as a prior issue of the best way to characterize the quantum theory or theories for which we are trying to draw out the ontological implications.

In this chapter, I will focus, as is standard and to keep the discussion manageable, on the interpretation of nonrelativistic quantum mechanics. In later chapters, we will extend our reach to relativistic quantum theories. To see the measurement problem and the ontological issues with which we will be primarily

The World in the Wave Function. Alyssa Ney, Oxford University Press (2021). © Oxford University Press.
DOI: 10.1093/oso/9780190097714.003.0001

concerned as clearly as possible, let's start with some basic facts about the mathematical tools used to represent simple as well as complex quantum systems.

One common such tool is the quantum wave function, ψ.[1] As Eugene Wigner put it, "Given any object, all the possible knowledge concerning that object can be given as its wave function" (1962, p. 173). The simplest wave functions describe states of a single particle in terms of a single attribute, for example, spin with respect to some particular axis, or momentum or position along a single dimension in physical space.

For example, we might consider the location of a single particle confined to move back and forth along one dimension (call it the x-axis) of a box of length a. We may then associate locations in the box with real numbers and represent possible position states of the particle using wave functions that take position values in, and yield numerical values out. Such wave functions may then be represented in the form ψ (x), since they are functions of the particle's position along the single x dimension of the box.

In general, wave functions may take a variety of forms. The particle in a box of length a has a wave function with the form:

$$\psi(x) = \sqrt{\frac{2}{a}}\sin\left(\frac{n\pi}{a}x\right),$$

where n may take on any positive integer value.[2] These values of n represent possible states the particle may be in, corresponding to different energy levels. The wave function of a particle in a box thus varies, depending on its energy state. But in all cases it is spread throughout the box, with different outputs at different locations

---

[1] For now, we will consider wave functions simply as mathematical tools used for representing quantum states. Later in the chapter, we will return to the question of what we should take the usefulness of these tools to suggest about the ontology of quantum theories.

[2] See, e.g., Griffiths (2005), p. 32.

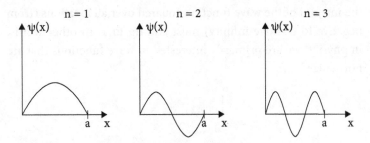

Figure 1.1. Wave Functions for the Particle in a Box

and can be graphed as shown in Figure 1.1 for the first three energy levels.

For the kind of idealized setup we are considering, in which a particle is truly confined, the wave functions will take on the value 0 at all locations outside of the box. In other words, $\psi(x < 0) = 0$ and $\psi(x > a) = 0$.

Noting that wave functions may be thought of as functions taking in precise values of some variable or other, such as position, raises the question of the significance of these values that are the outputs of wave functions. This is one of the main topics that will occupy us in this book: what in reality corresponds to these values that are the outputs of wave functions? These values are generally regarded as representing the amplitude of some kind of wave at various points in a space. This is indeed what is naturally suggested by the plots in Figure 1.1. But what is this wave? A genuine physical wave, one that is spread throughout the box in this example, or something else?

One rule built into quantum mechanics that is widely accepted and may guide us in discussion of these ontological issues concerns the connection between these values of the wave function and the results we should expect from measurements made on a system. It is commonly taken as a postulate of quantum mechanics[3] that

---

[3] See, e.g., Shankar (2012), p. 116.

the integral of the wave function squared over all locations (from negative to positive infinity) must add up to 1. In other words, in physics, we are primarily interested in wave functions that are normalized:

$$\int_{-\infty}^{\infty} |\psi(x)|^2 dx = 1.$$

This then allows us to associate the values of $|\psi(x)|^2$ with probabilities that upon measuring our system, we will find the particle at a given location. This is the Born rule. If we consider the region between any two specified locations x_1 and x_2, the Born rule states that we can calculate the probability that the system will be found in that region as $\int_{x_1}^{x_2} |\psi(x)|^2 dx$. We might then ask whether the proper way to interpret the wave function is not as a physical wave, but instead as something epistemic, as something related to the probabilities one ought to assign to measurement results. We will return to this question momentarily, however first let's note a few more important mathematical facts about wave functions.

From time to time in what follows, it will be useful to consider idealized situations in which a particle is known to be localized at one region or point in space. There are ways of using wave functions to represent states like this. For example, we could use the Dirac delta function, $\delta(x - a)$, where:

$$\delta(x) = \left\{ \begin{array}{ll} 0, & \text{if } x \neq 0 \\ \infty, & \text{if } x = 0 \end{array} \right\}, \quad \text{and} \quad \int_{-\infty}^{\infty} \delta(x) dx = 1.$$

The Dirac delta function $\delta(x - a)$ then represents the state of a particle using a wave function with an infinite output at point a and an output of 0 at all other locations.

However, a simpler and physically equivalent way of representing such states is the state vector notation.[4] For example, to represent a particle localized at x = 1, or at x = 2, we may write:

$$\psi_1 = |x = 1>,$$
$$\psi_2 = |x = 2>.[5]$$

Representing quantum states in this way more easily allows us to state what will be an important principle in what follows: the law of superposition. This is a basic principle of wave mechanics, built into quantum mechanics, which says simply that if ψ is a possible state and so is ψ', then $a\psi + b\psi'$ is also a possible state, where a and b are numbers. And so, for example, a third possible position state for our single particle will be:

$$\psi_3 = \frac{1}{\sqrt{2}} |x = 1> + \frac{1}{\sqrt{2}} |x = 2>.$$

This is actually not our first example of a superposition representation. In ψ_3, the system is represented as being in a superposition of two possible locations. But note that all of the quantum states described above (and plotted in Figure 1.1) for the particle in a box were also superpositions of the position variable, as the wave function took on nonzero values at more than one location in the box. For real quantum systems, superposition representations are ubiquitous.

When the state of a system is represented as a sum of terms using the state vector notation, to facilitate the connection between

---

[4] The state vector notation is generally most convenient and natural for representing states involving discrete-valued variables like spin, while the former are more convenient for systems taking on a continuous range of values like position and momentum. Because these notations are equivalent, state vectors will be frequently used to present wave functions.

[5] Here we use the Dirac ket notation in which quantum states are represented using angle brackets.

quantum states and probabilities encoded in the Born rule, quantum mechanics will again require that the representation of states be normalized.[6] To ensure this in state vector notation, the coefficients should be such that $|a|^2 + |b|^2 + |c|^2 + \ldots = 1$. We will then find that for a system represented by $\psi_1 = |x = 1>$, the Born rule entails that there is a probability 1 that the particle will be found at location $x = 1$ in the box upon measurement, and a probability 0 that the particle will be found anywhere else. For a system represented by $\psi_3 = \frac{1}{\sqrt{2}} |x = 1> + \frac{1}{\sqrt{2}} |x = 2>$, there is a probability ½ that the particle will be found at location $x = 1$ upon measurement, a probability ½ that the particle will found at location $x = 2$, and a probability 0 that the particle will be found anywhere else. When wave functions are normalized, the probabilities go by the square of the absolute values of the coefficients.[7]

Although it is common in physics to represent systems using wave functions in terms of a position variable as we just did, and as we will see, wave function realists and their opponents typically focus their discussions on position representation wave functions, in quantum mechanics, systems are often also represented in terms of other variables, such as momentum, energy, and spin.[8] And so

---

[6] The states should also be represented using an orthogonal basis (Shankar 2012, p. 9).

[7] This may be viewed as an alternative illustration of the claim that probabilities are associated with values of the wave function squared. If the vectors $|x=1>$ and $|x=2>$ are orthogonal, the inner products $<x=1|x=1> = <x=2|x=2> = 1$, and $<x=1|x=2> = <x=2|x=1> = 0$. The probability of finding a particle at $x=1$, then, can be found by multiplying the Dirac "ket" $\frac{1}{\sqrt{2}}|x=1> + \frac{1}{\sqrt{2}}|x=2>$, by the Dirac "bra" $<x=1|$, and squaring the result. This gives $\left[\frac{1}{\sqrt{2}} <x=1|x=1> + \frac{1}{\sqrt{2}} <x=1|x=2> \right]^2 = \left(\frac{1}{\sqrt{2}} \right)^2 = \frac{1}{2}$.

[8] In quantum mechanics, mass and charge are regarded as intrinsic properties of systems that do not vary. So wave functions will not be functions of mass or charge.

we have to keep in mind[9] that there are other ways of representing the states of quantum systems. We may use wave functions that are functions of position. But we also may use wave functions that are functions of momentum or other variables. There is a straightforward mathematical operation, the Fourier transform, that allows one to move back and forth between position and momentum representations. Indeed, one famous feature of quantum mechanics, the Heisenberg uncertainty principle, is essentially just a fact about the results of such Fourier transforming.

To see this, note that one way of writing down Heisenberg's 1927 uncertainty principle is:

$$\Delta x \Delta p \geq \frac{\hbar}{2},$$

where the deltas (Δ) represent uncertainties in the location or momentum of the system in question. So, for example, one may be more or less uncertain about the position (x) or momentum (p) of the particle in our box. Since the product of these uncertainties must take a minimum value of $\frac{\hbar}{2}$, this implies that the uncertainty in one value constrains the uncertainty in the other. And this may be seen as a consequence of the mathematical relationship between the shape wave functions take in a position representation of a system and the shape wave functions take in a momentum representation of the same system.

For example, consider the wave function for a free particle whose initial position is localized in the range –a < x < a. For very small values of a, that is where the particle is initially very well localized and so we have a low degree of uncertainty about its position, the

---

[9] This will be especially important in Chapter 4.

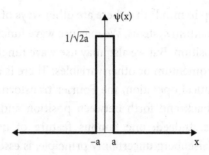

Figure 1.2. Wave Function for a Free Particle

wave function at that initial time may be written as $|\psi(x)| \approx \dfrac{1}{\sqrt{2a}}$, for values between a and −a, and ≈ 0 everywhere else.[10]

The corresponding function for the wave function squared may be written as $|\psi(x)|^2 \approx \dfrac{1}{2a}$ for values between a and −a, and ≈ 0 every where else, and thus the Born rule licenses our low degree of uncertainty about the particle's initial position. The probability that the particle is located at this initial time anywhere other than in the very narrow range between −a and a is effectively zero.

Using a Fourier transform, we may move from this position representation to a momentum representation of the particle in the same situation. The Fourier transform we need is given by:

$$\phi(k) = \frac{1}{\sqrt{2\pi}} \int_{-a}^{a} \psi(x)\, e^{-ikx} dx,$$

where k corresponds to the momentum of the particle,[11] and so $\phi(k)$ may be thought of as a momentum representation wave

---

[10] For more details, see Griffiths (2005), pp. 62–63.
[11] The de Broglie relation states that $p = \hbar k$.

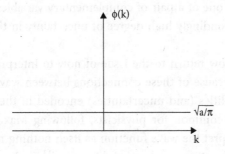

Figure 1.3. Momentum Representation for a Free Particle

function. Fourier transforming our position wave function will then yield:

$$\phi(k) = \frac{1}{\sqrt{a/\pi}} \frac{\sin ka}{k}$$

Using the small angle approximation, $\sin \theta \approx \theta$, we may see this function $\phi(k)$ as approximately equal to $\sqrt{\frac{a}{\pi}}$, for all values of k, plotting it as shown in Figure 1.3.

The momentum wave function of our particle is thus flat. And the shape of the corresponding function $|\phi(k)|^2 = \frac{a}{\pi}$ is similarly flat. From this, we infer using the Born rule that we should assign equal probability to the particle having any value of momentum. One should thus have maximal uncertainty about the momentum for a free particle that is initially localized between x = a and x = −a, that is one for which we have minimal uncertainty about its location. Uncertainties generally covary in this way. Following Niels Bohr (1928), we may thus call position and momentum complementary variables.[12] When there is a low degree of uncertainty in

---

[12] Other examples of complementary pairs of variables are energy and time duration and spins along orthogonal axes.

the value of one of a pair of complementary variables, there will be a correspondingly high degree of uncertainty in the value for the other.

We can now return to the issue of how to interpret the wave function. Because of these connections between wave functions and probabilities (and uncertainties) encoded in the Born rule, it has been common for physicists, following Max Born himself, to interpret the wave function as itself nothing more than a probability function, as a "probability wave."[13] For Born, the wave function does not directly represent something with objective (mind-independent) existence, like a physical wave. Rather, it is an epistemic item, something representing a subject's state of knowledge (or relative lack thereof) about the locations or other features of quantum systems. According to this point of view, particles always have definite positions and momenta at all times. It is just that the values of these positions and momenta cannot be simultaneously known. Wave functions summarize the limits of what can be known.

It is my view and the view of the majority of those who have worked to develop realist approaches to interpreting quantum theories that there exist good empirical reasons to reject the Born probability wave interpretation of the wave function. Let us be clear that to reject the Born probability wave interpretation of the wave function is not to reject the Born rule. The Born rule is only a statement about how wave function representations relate to the probabilities of measurement results. The Born interpretation says that wave functions (or the things wave function representations denote) *are* nothing more than summaries of the probabilities of measurement results.

---

[13] One finds this point of view expressed in Born's 1954 Nobel Prize address.

Experimental results demonstrating the wave behavior of quantum systems are the most obvious phenomena that undermine the Born probability interpretation of the wave function. The two-slit experiment, which has been carried out on systems as small as a single electron and as large as a molecule made up of 810 atoms, is the most famous such experiment.[14] It demonstrates that the wave function describes something that exhibits interference behavior, and so describes something objective and wave-like. This suggests that the uncertainties described by Heisenberg's principle capture something more than a mere epistemic limitation on our part. A natural thing to think on the basis of such results is that material systems (electrons and other objects built up out of them) themselves are the wave-like things directly represented by wave functions. Rather than being always confined to localized (perhaps point-like) positions as assumed by classical physics, electrons and molecules tend to be spread out over spatial locations like waves. Schrödinger used an analogy with clouds to describe the idea. To demonstrate the important distinction between the epistemic view of the wave function and the ontological one, Schrödinger noted: "There is a difference between a shaky and out-of-focus photograph and a snapshot of clouds and fog banks" (1935, p. 157). The Born interpretation would interpret the wave function as like the former. However, the interference phenomena revealed in the two-slit experiment and other results suggest something more like the latter. As we will see in the chapters that follow, ultimately the simple view that we should interpret wave functions as objective representations of individual (wave-like) quantum systems like electrons or molecules will not succeed. Electrons and molecules are not the waves described by quantum wave functions. But this does not

---

[14] Eibenberger et al. (2013).

undermine the empirical fact that there is something somewhere that is objectively wavelike and spread out, something directly captured by wave function representations.

Before we return to consider the measurement problem and how it relates to these ontological questions that are the main subject of this book, we will need one last element of the mathematical formalism making up nonrelativistic quantum mechanics. This is the description of how wave functions evolve over time: the time-dependent Schrödinger equation. So far, in our descriptions of the particle in a box and an initially localized free particle, the wave functions we have considered have just been functions of a single variable (x or p). However, the wave functions used to describe systems' dynamical evolution will be functions of time as well as position or other state variables. The Schrödinger equation may then be stated simply in the form:

$$\hat{H}\,\psi = i\hbar\frac{\partial\psi}{\partial t},$$

where "$\hbar$" is Planck's constant, "i" is $\sqrt{-1}$, and "$\hat{H}$" is the Hamiltonian, a way of writing down a system's total energy. For simple systems of the kind we have been considering, we may expand the Hamiltonian into a sum of kinetic and potential energy terms, writing the Schrödinger equation instead as:

$$-\frac{\hbar^2}{2m}\frac{\partial^2\psi}{\partial x^2} + V\psi = i\hbar\frac{\partial\psi}{\partial t}$$

This equation has several features that will be relevant in what follows.

First, the Schrödinger equation is linear. This means that if the law takes a solution A into a solution A' at a later time, then it will take a solution cA into a solution cA' at that later time. And also, if the law takes a solution A into the solution A' and another solution B into the solution B', then it will take the solution A + B into the solution A' + B'. This latter fact ensures that if one starts with a state that is spread out over different values of (e.g.) position, such as the state ψ_3 we considered above:

$$\psi_3 = \frac{1}{\sqrt{2}}\psi_1 + \frac{1}{\sqrt{2}}\psi_2,$$

and ψ_1 evolves into ψ_1' and ψ_2 evolves into ψ_2', then this state will not evolve at a later time into a state in which one of the ψ_1' or ψ_2' terms drops away. It will necessarily evolve to a later state of the form:

$$\psi_3' = \frac{1}{\sqrt{2}}\psi_1' + \frac{1}{\sqrt{2}}\psi_2'.$$

The system will maintain its superposition of position.

A second fact about the time-dependent Schrödinger equation is that it is unitary. Essentially what this ensures is that time evolution preserves the normalization of the wave function. If a wave function that is a solution to the Schrödinger equation is normalized, so that $\int_{-\infty}^{\infty} |\psi(x)|^2 dx = 1$, then the time-evolved wave function will also be so normalized.

Finally, the time-dependent Schrödinger equation is deterministic. If one knows the wave function of a system at a given time, then if one evolves this wave function forward in time using the Schrödinger equation, one will know with certainty the wave function of the system at any later time. The Schrödinger equation is a law that determines later states of systems; it is not a probabilistic law. And so, if God had written that the central law governing the universe was to be the time-dependent Schrödinger equation, then, to paraphrase Einstein, God would not have been playing dice.[15]

1.2 The Measurement Problem

We now have what we need in order to see the measurement problem for quantum mechanics and what are the typical means of solving it. The measurement problem may be roughly stated as the problem of how it is that quantum systems may evolve from states with wave functions that are spread out, states that are superpositions with respect to a variable, into the sort of determinate, localized states we see upon measurement. Quantum mechanics allows that systems may evolve into states that are superpositions of one variable or another, of position or momentum or spin. And yet, when we make measurements to try to detect a system at one location or another, or do an experiment to measure a system's momentum or energy or spin, we always find the system to possess one determinate value or other. So, somehow systems seem to be evolving from states whose wave functions are spread out, states that are indeterminate with respect to the value of a given variable, into states that are determinate with respect to that variable.

But this seems impossible given what we have just said about the character of the law of quantum mechanics that describes how systems evolve over time, the Schrödinger equation. First, the Schrödinger equation is a deterministic law. This means that if a system is in one state at a given time, then that law determines the states the system will be in at all later times. And second, we have seen that the Schrödinger equation is linear. So if an initial state is a superposition of A and B states, and A evolves to A' and B evolves

---

[15] Born and Einstein (2005), p. 149. As we will shortly see, it is the controversial collapse postulate that brings objective chance and indeterminism into what would otherwise be the deterministic framework of quantum mechanics. See also Lewis (2016), Chapter 6.

to B', then the later state will be a superposition of A' and B' states. If the earlier state's wave function is spread out, then the later state's wave function will also be spread out.[16]

Thus, facts about quantum states and their evolution appear to conflict with what we observe as the results of our measurements on quantum systems. We may thus view the measurement problem in the form of an inconsistent triad:

(1) Quantum mechanics allows systems to evolve into states that are spread out with respect to variables such as their positions.

(2) Once systems evolve into states that are spread out with respect to one or another variable, quantum mechanics ensures they will continue to be spread out into the future.

(3) Measurements always produce results that show systems are localized or determinate with respect to the values of variables.

If we want a version of quantum mechanics that is compatible with what we observe when we conduct measurements, we will have to either change how quantum mechanics describes the states systems may enter into by rejecting (1), change how quantum mechanics says systems evolve (by rejecting the Schrödinger equation or saying it fails to obtain at least some of time), thereby rejecting (2), or give up the claim that measurements always have localized or determinate results, even though they seem to, rejecting (3).

1.3 Orthodox Quantum Mechanics

Bohr (1928) seems to have thought that he had solved this problem by appealing to the complementarity of different

---

[16] For further discussion, see Albert (1992), Chapter 4.

forms of description, such complementarity as earlier had been demonstrated to obtain between variables related by the Heisenberg uncertainty relations. For Bohr, complementarity entails that when a system is correctly described as possessing one feature, the attribution of complementary modes of description are inappropriate.

Bohr notes it is true that systems tend to evolve into states that may be usefully represented as superpositions of one or another variable. For example, the wave function for a system written down in terms of position may be spread out over a range of locations. Yet somehow making a measurement of the position of such a system brings about a situation which we would no longer describe as being a superposition. But what Bohr calls the quantum postulate implies that this should not be thought of as the system's undergoing some kind of evolution from a state that is spread out to one that is localized. For just as position and momentum representations are complementary, so are quantum mechanical and observer-based descriptions:

> On one hand, the definition of the state of a physical system, as ordinarily understood, claims the elimination of all external disturbances. But in that case, according to the quantum postulate, any observation will be impossible, and above all, the concepts of space and time lose their immediate sense. On the other hand, if in order to make observation possible we permit certain interactions with suitable agencies of measurement, not belonging to the system, an unambiguous definition of the state of the system is no longer possible, and there can be no question of causality in the ordinary sense of the word. The very nature of quantum theory thus forces us to regard the space-time co-ordinatization and the claim of causality, the union of which characterizes the classical theories, as complementary but exclusive features of the description, symbolizing the idealization of observation and definition respectively. (Bohr 1928, p. 580)

This emphasis on the complementarity of causal and spatiotemporal descriptions, and of quantum mechanical and observer-based descriptions in Bohr's account, and its use in rejecting a quantum description of measurement, is one point of view that is sometimes referred to as the Copenhagen interpretation of quantum mechanics.[17] If one adopts this point of view, then one should not look for a resolution of the measurement problem in terms of an account of how systems evolve from indeterminacy into determinacy using quantum physics. But then one may ask why we should not demand more out of our quantum theories, particularly if they are to provide for us fundamental understanding of the states and behavior of microscopic systems.

Indeed, some of Bohr's colleagues wanted more out of quantum mechanics, proposing modifications to its mathematical formalism in order to accommodate a description of what takes place during measurement. One influential such proposal, introducing the so-called collapse of the wave function, was presented in the work of John von Neumann.[18] In Von Neumann's 1932 book *The*

---

[17] See Becker (2018) for an examination of the ambiguities in what is generally regarded as *the* Copenhagen interpretation.

[18] Heisenberg, in his 1927 paper introducing the uncertainty principle, discusses measurements as involving "quantum jumps" (p. 22), stochastic processes which may be regarded as collapsing the wave function. But it is clear that Heisenberg is not intending to provide a physical description of a causal process in nature, as opposed to a manner in which we may move from relative uncertainty about the state of a system to relative certainty:

> In the rigorous formulation of the law of causality—"If we know the present precisely, we can calculate the future"—it is not the conclusion that is faulty, but the premise. We simply can not know the present in principle in all of its parameters. Therefore all perception is a selection from a totality of possibilities and a limitation of what is possible in the future. Since the statistical nature of quantum theory is so closely tied to the uncertainty in all observations or perceptions, one could be tempted to conclude that behind the observed, statistical world a "real" world is hidden, in which the law of causality is applicable. We want to state explicitly that we believe such speculations to be both fruitless and pointless. The only task of physics is to describe the relation between observations. (Heisenberg 1927, p. 32)

Heisenberg, like Bohr, rejects the desire to provide a quantum theory that includes a description of the physical processes leading to the measurement outcome.

Mathematical Foundations of Quantum Mechanics, he proposes we view quantum mechanics as describing the existence of two kinds of process that operate in nature: the "arbitrary changes [in systems brought about] by measurements" and "the automatic changes which occur with passage of time" (1932, p. 351). Von Neumann labels the former as Process 1. These are stochastic processes. They are those in which, for the case of systems initially in a superposition of the variable to be measured, there are several possible later states into which these systems may evolve. The probabilities of evolution into one state or another are given by the wave function of the system in accordance with the Born rule. Process 1 is thus understood as one of state reduction or collapse of the wave function, from a state that is spread out over several values of the variable being measured, to one of localization onto a single determinate value of that variable. By contrast, Von Neumann's Process 2 takes place when a measurement is not being made on a system. This is unitary evolution according to the Schrödinger equation, which as we have seen, allows wave functions to continue to spread out and constitutes a deterministic process.

Von Neumann thus solves the measurement problem by rejecting what I above labeled Claim (2):

(2) Once systems evolve into states that are spread out with respect to one or another variable, quantum mechanics ensures they will continue to be spread out into the future.

By postulating a primitive process of measurement or Process 1, von Neumann thereby makes room in quantum mechanics for the evolution of systems into localized states.

Von Neumann's interpretation of the quantum formalism as involving two kinds of processes has been so extremely influential as to become what commonly appears in textbooks. For this reason, it is often referred to as the "orthodox" or "textbook" formulation of (nonrelativistic) quantum mechanics.[19] Many believe, based on this formulation, that the quantum world involves, at least some of the time, when systems interact with observers and measuring devices, objectively stochastic processes and thus, quantum mechanics has proven determinism wrong. Yet those working in the foundations of quantum mechanics have largely been dissatisfied with this orthodox formulation. And so we should not be so hasty in drawing the conclusion that quantum mechanics is incompatible with determinism.

Probably the best (at least the most colorful and famous) statement of the problems raised by the orthodox formulation of quantum mechanics was given by the physicist John Bell in a paper from 1989, "Against 'Measurement.'" In that paper, Bell pleaded for a reformulation of quantum mechanics as a theory that did not use the concept of measurement in the statement of its basic laws and processes. As we will see in Section 1.4, there are indeed several ways of solving the measurement problem that avoid the postulation of a primitive process of measurement.

Bell and others have outlined several problems with including a primitive concept of measurement in the formulation of quantum theories. First, quantum theories aspire to give precise, scientific accounts of physical phenomena. As such, they should not make use of the kind of imprecise or vague language that pervades ordinary speech. But 'measurement' is an extraordinarily *imprecise* term. There is nothing like a precise meaning associated with the word that would tell us in exactly which circumstances a measurement occurs and in exactly which circumstances it does not. As with many vague terms ('bald,' 'tall,' 'game'), there are clear settings

---

[19] For example, Griffiths, in the standard textbook for undergraduate courses in quantum mechanics stipulates: "Upon measurement, the wave function 'collapses' to the corresponding eigenstate" (2005, p. 106). Shankar (2012), a very popular graduate textbook, lists this fact about wave function collapse as the third postulate of quantum mechanics (p. 116).

in which we know the concept applies and clear settings in which we know it does not, but there are also cases in which the matter is not so clear. Wittgenstein (1953) noted the ambiguity in 'game.' Baseball is a game. Carpentry is not. But whether hunting is may be a matter of debate. Likewise, there are cases in which a scientist is clearly performing a measurement, and clear cases when one is not. But there are other sorts of cases, for example, interactions between microscopic quantum systems and observers taking place outside of a lab, for which we can raise the question: is a measurement taking place or is it not? There must be an answer if there is to be a precise fact about which von Neumann process is taking place.

Even if we could solve this problem and settle which sorts of situations are cases of measurement, for any individual such process, if one thinks carefully about the details, one realizes there is no precise answer to when and where the measurement began and when and where it ended, there being a continuous series of events from experimental interaction to conscious realization of a result. And so again, there fails to be an answer as to the precise circumstances in which von Neumann's two processes operate. Thus the textbook formulation of quantum mechanics fails to have the kind of precision we should expect from our best physical theories.

To further illustrate the concern, suppose we try to measure the spin of an atom using a Stern-Gerlach apparatus. Such a device works by directing a source of atoms through a pair of magnets that may deflect them up or down based on their spin and toward a screen on which the atoms' impact may be detected (see Figure 1.4).

At what point in this chain of events, we may ask, does the "measurement" take place? Does it occur at the instant and location when the atoms stop heading in a straight line and are instead deflected up or down by the magnets? Or does it take place only after they make their marks on the screen? Or has the measurement

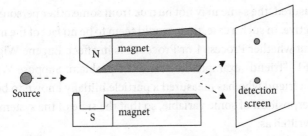

Figure 1.4. Stern-Gerlach Device

not actually taken place until the scientist first sees a mark on the detection screen? Or when she forms the conscious belief that such a detection has taken place? There seems no clear answer to these questions.

In fact the trouble with the imprecision of 'measurement' is so severe that it prompted Bell to ask:

> Was the wavefunction of the world waiting to jump for thousands of millions of years until a single-celled living creature appeared? Or did it have to wait a little longer, for some better qualified system . . . with a Ph.D.? If the theory is to apply to anything but highly idealized laboratory operations, are we not obliged to admit that more or less "measurement-like" processes are going on more or less all the time, more or less everywhere? (p. 216)

Without an answer to these questions, an answer that simply seems blocked by the imprecision of the concept of measurement, we cannot know how to apply von Neumann's proposed formulation of quantum mechanics to real circumstances.

That is the first problem. Another that many, including Bell, have raised is that measurement is a *relative* concept. While from one person's perspective, the value of a certain variable may have been measured, the same may not be true from some other person's perspective. In such a case, there would seem to be no fact of the matter about whether Process 1 or Process 2 is in effect. Eugene Wigner's (1961) "friend" example illustrates the problem. Suppose Wigner has a friend who has measured a particle initially known to be in a superposition of some variable, so that the state of the system may be written as:

$$\psi_P = a\,|A>_P + b\,|B>_P,$$

where subscripts are used to highlight that these kets refer to the state of the particle "P." Wigner wants to represent the state of his friend. Because he does not know which result his friend has found, relative to Wigner, a measurement has not yet been made on the system. This means that to represent the state of his friend, Wigner will write down a wave function that describes the combined state of the system and his friend as:

$$\psi_{PF} = a'\,|A>_P|\ \text{"A is the result"}>_F + b'\,|B>_P|\ \text{"B is the result"}>_F$$

Wigner can then measure this combined particle + friend system by asking his friend which result he found. According to the orthodox formulation, this measurement will collapse the combined system onto one of the two states:

$$|A>_P|\ \text{"A is the result"}>_F$$

Or

$$|B>_P|\ \text{"B is the result"}>_F.$$

By the Born rule, the probabilities that Wigner will get one or another result are $|a'|^2$ or $|b'|^2$, respectively. Wigner notes that this is what orthodox quantum mechanics appears to suggest, since it states that it is measurements that collapse wave functions. But, he notes, this "appears absurd because it implies that my friend was in a state of suspended animation before he answered my question" (1962, p. 180).

Wigner's proposed solution to this problem is to insist that conscious agents don't enter into superpositions like ψ_{PF}. And so, given that his friend is conscious, the state of the combined system including the friend will be determinate. This means there is an absolute fact about when collapse of the wave function occurred, some fact determining the state of his friend that does not vary from observer to observer. This is a fact that does not depend on whether Wigner measures the combined system or not, thus avoiding absurdity.

Wigner takes the facts about when systems involve conscious agents to be primitive, not capable of analysis in physical (or more specifically, quantum mechanical) terms. He accepts without qualms that this approach commits him to a dualistic view about the mind-body problem, that facts about consciousness are primitive and do not reduce to any other physical facts about systems. But this is what he thinks a precise and objective account of how quantum systems evolve into determinate states requires. For Wigner, quantum mechanics teaches us something important about the relationship between minds and bodies.

Wigner thus provides a way of modifying the orthodox formulation of quantum mechanics so that it seemingly avoids the problems of imprecision and relativity arising from formulations couched in terms of measurement. However, even if we accept that the facts about consciousness are precise and absolute in a way that pre-theoretic facts about measurement are not, there is still a third problem with the orthodox formulation that even Wigner's proposal does not resolve. This is that whether we stipulate that the wave functions of systems collapse upon measurement or interaction with a conscious agent, for systems to take on determinate values of some variable, these formulations require that there must always be something outside of a system: something that makes the measurement, or something conscious that interacts with the system (see also Everett 1957). But this seems problematic if we would like quantum theories eventually to apply to everything there is, including the universe or cosmos as a whole. For assuming the whole cosmos evolves into something describable as a superposition with respect to some variable, these formulations allow no way for it evolve out of such a state. It seems then that the textbook formulation must imply that the state of the cosmos is always indeterminate, since there will never be anything outside of it to measure it. And then what is to be said for the state of all of its subsystems, all of the smaller things inside?

In sum, due to all of these problems, physicists and philosophers have sought out solutions to the measurement problem that reformulate the basic principles of quantum mechanics without making use of the concept of measurement, or of consciousness. This has been done in a variety of ways, with proposals that alternatively appeal to hidden (or additional) variables, spontaneous (i.e., non-measurement-induced) collapses, and many worlds or relative states. In principle, the framework developed in this book, wave function realism, is compatible with any of these approaches to solving the measurement problem, including even (what we've just seen as problematic) orthodox quantum mechanics with collapse upon measurement. In my view, wave function realism is a framework that combines more naturally with some approaches to solving the measurement problem than it does with others. As we will see, I don't find wave function realism to be especially well motivated as an ontological interpretation of hidden variables theories. But wave function realists have worked within the context of all of these major approaches to solving the measurement problem, all of these ways of attempting to achieve a precise statement of quantum mechanics. And so I will say a bit more about each such approach in the next section.

1.4 Quantum Mechanics without Measurement

In this book, I have no desire to enter the debate about which of the following is the best approach to solving the measurement problem for quantum mechanics. And so although I will be stating each of the three main approaches to solving the measurement problem so that we can see how they combine with wave function realism and other attempts at answering the ontological questions, I will not attempt an exhaustive reckoning of the points for and against each of them.[20] I think it will be clear over the course of this book that I am most sympathetic to the Everettian framework, especially because it applies most straightforwardly to relativistic quantum theories like quantum field theories and theories of quantum gravity. But the Everettian approach does face its own problems (most notably the problem of how to make sense of the statistical claims that form our evidence for quantum mechanics)[21] and so it is worth leaving all options on the table.

I will describe the three main ways in which physicists have approached the question of how to understand quantum mechanics without invoking basic principles about measurement or consciousness. These are:

to reinterpret the sort of states quantum systems can be in,
to reinterpret the way quantum mechanics says quantum states evolve, and

---

[20] Albert's (1992) evaluates the different approaches. Wallace's (2012) defends the Everett approach. Dürr, Goldstein, and Zanghì (2013) defend Bohmian mechanics.
[21] See Lewis (2016), Chapter 6. But see also Wallace (2012) and, more recently, Carroll (2019).

to reject the claim that measurements always produce determinate results.

(Orthodox quantum mechanics, it is easily seen, is a version of
the second strategy.) I call these "approaches" to solving the measurement problem because, as we will see, each may be applied to
generate a variety of more specific quantum theories. Each is an
approach to reconciling quantum physics with the fact that our
measurements seem to produce determinate results. Each rejects
one of the three principles that (as noted in Section 1.2) together
appear to form an inconsistent triad.[22]

The first approach involves reinterpreting the sort of states
quantum systems may enter into. If we were initially concerned
about how quantum systems may evolve (given Schrödinger evolution) from indeterminacy into the kind of determinacy we see
upon measurement, then this problem goes away if we deny that
quantum systems ever enter into indeterminacy in the first place.
This is often labeled a hidden variables strategy to solving the
measurement problem as the idea is to supplement wave function
representations with facts about determinate values that systems
take on for certain variables (and laws about these values' evolution). The values for these variables are "hidden" in the sense that
they are not described by standard wave function presentations of
quantum mechanics such as that outlined in Section 1.1.

The hidden variables theory most commonly defended today[23] is
Bohmian mechanics, which follows a strategy for solving the measurement problem offered by David Bohm in 1952. It's in many ways
similar to a proposal of De Broglie from 1927, and so it's sometimes
referred to as the De Broglie-Bohm theory.[24] Bohmians propose that

---

[22] See Sebens (2015) for an approach that combines aspects of the first and the third
strategies.

[23] Modal interpretations provide other examples of hidden variables theories, e.g., van
Fraassen (1991).

[24] This was presented at the 1927 Solvay conference.

quantum particles always possess definite positions, even when their wave functions are spread out over multiple values. So for Bohmians, positions are the hidden variables. Bohmians interpret the experimental evidence suggesting that quantum systems are wave-like as evidence that there is *something* that should be described by a wave equation like the Schrödinger equation. This is the quantum wave function.[25] But for Bohmians, the wave function is distinct from the material systems we measure and otherwise interact with.

Since Bohmians add facts about determinate positions to facts about the wave function, they must also add a law to their formulation of quantum mechanics to describe how these positions evolve over time. One way of doing so is to supplement the Schrödinger equation (the law describing the evolution of the wave function) with a so-called guidance equation, as in Dürr, Goldstein, and Zanghì (1992).[26] Where Q_k refers to a position degree of freedom at a time, change in Q_k is given by:

$$\frac{dQ_k}{dt} = \left(\frac{\hbar}{m_k} \right) \operatorname{Im} \frac{\Psi^* \partial_k \Psi}{\Psi^* \Psi} (Q_1, \ldots, Q_N).$$

The guidance equation, another deterministic law, thus shows how the evolution of determinate positions is guided by the state of the wave function.

The Bohmian argues that measurements always receive determinate results because quantum systems always possess determinate positions. Measurements are always encoded in the position of something—be it the tip of a pointer, a dot on a screen, a mark in a notepad (cf. Bell 1982, Albert 1992). To understand how these positions evolve over time, we must make use of wave function

---

[25] Here 'wave function' is being used to denote something in the world, perhaps concrete, something other than the mathematical tool used to get at this ontology.

[26] Dürr, Goldstein, and Zanghì's approach to developing Bohm's theory differs in substantial ways from others. See also Bohm and Hiley (1993).

representations, but these wave function representations do not imply any indeterminacy in position. Bohmians thus reject what I earlier referred to as Claim (1):

(1) Quantum mechanics allows systems to evolve into states that are spread out with respect to variables such as their positions.

According to Bohmian mechanics, quantum systems themselves (electrons, molecules, etc.) are never spread out or indeterminate with respect to their positions, even if wave functions are.

A second approach to solving the measurement problem involves modifying the evolution of the wave function so that at least some of the time, systems don't evolve in accordance with the Schrödinger equation, but may evolve out of states that are indeterminate with respect to one variable and into a state that has a determinate value of the kind we see upon measurement. As we've seen, this involves the rejection of Claim (2):

(2) Once systems evolve into states that are spread out with respect to one or another variable, quantum mechanics ensures they will continue to be spread out into the future.

For the reasons canvassed in Section 1.3, it would be nice to have an account of how this works that does not simply say wave functions collapse when one makes a measurement on a system. Philip Pearle (1976), as well as Giancarlo Ghirardi, Alberto Rimini, and Tullio Weber (1986), have proposed one way of accomplishing this.[27] The central idea is to say that collapses are not triggered by measurements but instead occur spontaneously. For simple systems of, for example, a single particle, there is a small but finite probability that the system's wave function will undergo a hit or collapse.

---

[27] Bell (1987) gives a nice overview of this approach.

This is modeled mathematically by the multiplication of its wave function at a time by a Gaussian function with a certain width.[28]

The probability of collapse for a simple and isolated system is so small that we should not expect such a collapse to occur in our lifetimes. However, when a particle interacts with a large group of such particles, as will be the case in, for example, the kind of interaction we would ordinarily think of as a measurement, one in which the particle is itself one component of a larger macroscopic system, there are so many such simple systems present ($\sim 10^{23}$) that the probability of one among the group undergoing a hit becomes very large. This then triggers a collapse of the whole system.[29] The probability of this hit localizing around one or another value is determined by the wave function of the system in accordance with the Born rule. After a hit, the system returns to unitary evolution in accordance with the Schrödinger equation until another hit occurs. This is how and why systems whose wave functions are initially

---

[28] The width of this function and the probability of collapse at a time for a simple system are both new constants proposed as part of the theory whose values may be (and have been) tested experimentally.

[29] To see why, we may return to Wigner's friend scenario. We earlier represented the particle + friend state as: $\psi_{PF} = a'|A>_p|$ "A is the result" $>_F + b'|B>_p|$ "B is the result" $>_F$. But now notice that $|$ "A is the result" $>_F$ and $|$ "B is the result" $>_F$ each summarize complicated products (or sums of products, but set this aside) of many individual particle states $|A_2>_{P2} \cdots |A_n>_{Pn}$ and $|B_2>_{P2} \cdots |B_n>_{Pn}|$, with each $|A_k>_{Pk}$ or $|B_k>_{Pk}$ representing the location of an individual particle making up Wigner's friend. So the combined particle + friend state may be rewritten as: $\psi_{PF} = a'|A_1>_{P1} |A_2>_{P2} \cdots |A_n>_{Pn} + b'|B_1>_{P1}| B_2>_{P2} \cdots | B_n>_{Pn}$. One can thus see that there are only two possible states the system may collapse into:

$$|A_1 >_{P1} \ldots |A_n >_{Pn}, \quad \text{with probability} |a'|^2$$

and

$$|B_1 >_{P1} \ldots |B_n >_{Pn}, \quad \text{with probability} |b'|^2.$$

If an individual particle collapses into one of the A states, this will trigger all of the particles to collapse onto an A state. And if an individual particle collapses into one of the B states, this will trigger all of the particles to collapse onto an B state. This is implied by the Born rule, since the probability of finding any mixed A/B state upon measurement, given initial state ψ_{PF}, is zero.

quite spread out may evolve to have determinate values upon measurement. There are several different forms of spontaneous collapse theories whose exact predictions for how and when wave function hits occur are not identical.[30] But in what follows, we won't need to concern ourselves with these disagreements.

So far, we have seen ways of responding to the measurement problem by (a) modifying the standard account's view about the kinds of states quantum systems may enter into and by (b) modifying the rules of state evolution. The third way physicists have proposed to solve the measurement problem is by (c) rejecting what I have called Claim (3), that:

(3) Measurements always produce results that show systems are localized or determinate with respect to the values of variables.

This approach follows a proposal by the physicist Hugh Everett III in his 1957 doctoral dissertation. It is what is now referred to as the many worlds interpretation of quantum mechanics; however, Everett called it the theory of relative states. The core idea is to allow that systems as a whole may evolve into states that are spread out among several values and that these systems' wave functions never collapse onto single values. Rather, what measurements do is correlate observers' experiences with parts of systems' total states, that is, with relative states involving determinate values. And so that is why measurements appear to have determinate results, even though systems' wave functions remain spread out.

To illustrate, consider an idealized case in which a particle has evolved into the following state that is indeterminate with respect to its position:

---

[30] For example, a successor theory, CSL (the continuous spontaneous localization model), was proposed by Ghirardi, Pearle, and Rimini (1990). Roger Penrose has developed an alternative spontaneous collapse model (1989) that ties wave function collapse to facts about spacetime geometry.

$$\psi = \frac{1}{\sqrt{2}} \,|x = 1> + \frac{1}{\sqrt{2}} \,|x = 3>$$

Suppose we want to know how it can be that when we try to measure or locate the particle, we find that it isn't spread out over the two locations, x = 1 and x = 3, but is rather always found at one of them, x = 1 or x = 3. As we saw, the defender of a hidden variables interpretation will say that although there really is something wave-like and spread out, the particle is not, since position values of quantum systems are always determinate. Spontaneous collapse theorists will say that when a measurement occurs, the system interacts with the very many particles in the measuring device, and a collapse in one of these particles' wave functions triggers a collapse in the combined particle-measuring device system, making the initial particle take on a determinate position.

The Everettian rather will say that interaction with the measuring device causes the system to evolve into a more complex state we can represent as:

$$\psi' = \frac{1}{\sqrt{2}} \,|x = 1>_P \,|\text{``1''}>_M + \frac{1}{\sqrt{2}} \,|x = 3>_P \,|\text{``3''}>_M.$$

This is not a newly postulated quantum process, but just what one would predict assuming standard Schrödinger evolution. I've introduced subscripts to show what is the part of the representation referring to the particle and its position variable (P) and what is the part referring to the measuring device and its reading (M). In ψ', we see a representation of a more complex system involving the particle and measuring device, where facts about the total system are still indeterminate. The Everettian now wants us to take this one step further and consider this complex system's interaction with an observer (O) who may now view the reading on the measuring device and form a belief about which result has been recorded. This will produce a state we may write down as:

$$\psi'' = \frac{1}{\sqrt{2}} |x = 1>_P |``1" >_M| ``\text{Result is } 1">_O$$
$$+ \frac{1}{\sqrt{2}} |x = 3>_P |``3" >_M| ``\text{Result is } 3">_O$$

Because the Schrödinger equation does not allow wave function collapse, as the observer interacts with the system, she too will become part of a total quantum state that is spread out over two possible measurement results.[31] And so, the Everettian rejects what we above called Claim (3), that measurements always produce determinate results for the values of variables. Instead, the Everettian allows that the quantum state remains distributed among different possible values, even after measurement. The claim is that this is actually compatible with our experiences of measurements, our sense that measurements *seem* to take on determinate results. When you as an experimenter make a measurement and see a determinate result, that constitutes a relative state, and there is another relative state in which you find another determinate result. These distinct relative parts do not interact, and so it won't seem to you that there is any indeterminacy.[32] Thus the quantum formalism is made compatible with the fact that measurements seem to always produce determinate results.

Although Everett did not use this terminology when presenting his solution to the measurement problem, later proponents have

[31] This is of course similar to what we went through earlier in discussion of the Wigner friend scenario. Wigner argued that the result was absurd since states of consciousness cannot be so indeterminate, but Everett asked us to go ahead and see how we can make sense of states like this.

[32] The explanation of why the relative states do not interact became better understood after the publication of Everett's paper in the 1950s due to work on the role of decoherence in quantum mechanics (Barrett and Byrne 2012, pp. 49–50).

thought of this as a many worlds theory.[33] In addition to the world we observe, that which surrounds us and with which we interact, there are many other worlds, other parts of the total quantum state. These are what are represented by the other terms in the superposition of the wave function for our universe. This consequence is certainly surprising, but for many it is exhilarating to think that quantum mechanics may be telling us that there exist so many other worlds we never observe.[34]

1.5 Wave Function Realism

I started this chapter by noting that the central questions with which this book is concerned are ontological, and that these questions should not be conflated with a distinct question, that of the correct solution to the measurement problem. I hope it is now clear that these topics are distinct. The measurement problem forces us to ask whether the standard formulation of quantum mechanics may be interpreted in such a way so as to be consistent with our experiences (as the Everettian thinks), or whether wave function representations must be supplemented with the postulation of variables with determinate values (as the Bohmian argues), or whether we should modify the way quantum mechanics represents the evolution of states (as the collapse theorist suggests). This gives us three distinct kinds of quantum formalisms. Table 1.1 summarizes the situation for nonrelativistic quantum mechanics.

The ontological questions then ask us what kinds of worlds we should take these versions of quantum mechanics to be describing.

---

[33] The first to use this language appears to have been Bryce Dewitt (1970). Christina Conroy (2011) argues that we should resist this talk of many worlds and proposes an alternative interpretation of Everett's "relative states."

[34] Sean Carroll calls it the "courageous" formulation of quantum mechanics (2019).

Table 1.1 Three Versions of Quantum Mechanics Answering the Measurement Problem

Measurement Problem Solution	State Representation	Law(s) of State Evolution
Everettian quantum mechanics	Wave function	Schrödinger equation
Bohmian mechanics	Wave function + Position coordinates	Schrödinger equation Guidance equation
Collapse theories	Wave function	Stochastic GRW equation

Different strategies for solving the measurement problem may correspond to very different pictures of objective reality.

One may begin addressing the ontological questions by noting that each interpretation of the quantum formalism makes central use of a wave function representation and includes a law or laws describing the evolution of this wave function over time, and possibly how the state of this wave function affects the evolution of other basic entities. Literally understood, this suggests there is a physical entity these wave function representations are describing. It is standard today to refer to this entity using the same name as the mathematical tool used to represent it.[35] The wave function realist understands this entity as a wave or a field, because its evolution is described by a wave equation and, like other fields, is defined at any moment by a range of amplitude values it takes on at each point in its space.[36] As Albert writes:

> The sorts of physical objects that wave functions *are*, on this way of thinking, are (plainly) *fields*—which is to say that they are the

[35] The context should always make it clear whether I am using the phrase "the wave function" to refer to the physical entity or to the mathematical object used to represent it.
[36] It also carries values of phase, but this will not concern us at this point in the discussion.

sorts of objects whose states one specifies by specifying the values of some set of numbers at every point in the space where they live, the sorts of objects whose states one specifies (in *this* case) by specifying the values of *two* numbers (one of which is usually referred to as an amplitude, and the other as a *phase*) at every point in the universe's so-called *configuration* space.[37] (1996, p. 278)

One might then think that the different formulations of quantum mechanics that solve the measurement problem each say different things about the wave function, either about how it evolves or whether it is the unique basic element of a quantum ontology. But in each case, since wave function representations are successfully used to make predictions, it is natural to take such representations literally, to correctly describe objective facts about this entity, rather than facts about something else.

One finds this form of reasoning particularly explicit in a 2004 paper by Peter Lewis:

> The wavefunction figures in quantum mechanics in much the same way that particle configurations figure in classical mechanics; its evolution over time successfully explains our observations. So absent some compelling argument to the contrary, the prima facie conclusion is that the wavefunction should be accorded the same status that we used to accord to particle configurations. Realists, then, should regard the wavefunction as part of the basic furniture of the world. . . . This conclusion is independent of the theoretical choices one might make in response to the measurement problem. . . . [I]t is the wavefunction that plays the central explanatory and predictive role. (2004, p. 714)

The prima facie case then stems from the ubiquity and success of wave function representations in quantum mechanics and

---

[37] More on "configuration space" in a moment.

the naturalness of identifying wave functions with the physical fields these representations describe. Just as the success of classical physics previously committed us to particle configurations, quantum physics now commits us to wave functions.

In Chapter 2, I would like to see what we can say about strengthening this prima facie case into an all-things-considered or ultima facie case for wave function realism. As I mentioned in the Preface, Albert (and others as well) have represented wave function realism not merely as an approach one *could* adopt to interpreting quantum mechanics initially suggested by the formalism, but as the *only* tenable way to interpret quantum mechanics. But first I want to take issue with a claim that both appears in this passage from Lewis and is commonly made by wave function realists (Albert included). This is the claim that wave function realism is an approach to providing an ontological interpretation of quantum mechanics that receives equal support however one chooses to solve the measurement problem, that as Lewis puts it, this issue "is independent of the theoretical choices one might make in response to the measurement problem." This will also help us better understand the picture of the world the wave function realist is offering.

Each way of formulating quantum mechanics so as to solve the measurement problem makes central use of wave function representations. And so whichever approach we adopt to solving the measurement problem, there is prima facie reason to take wave functions to be representing *something* that is real.[38] But this is weaker than what is required by wave function realism. For wave function realism is the view that wave functions are not just real, but are real, objective, physical fields. Although it is natural to take this attitude toward wave functions as they are presented in Everettian and collapse approaches, this is not similarly so

[38] In fact, as I noted earlier, there is more than a prima facie case for this claim as the experimental evidence for quantum mechanics is experimental evidence for something wave-like in the behavior of matter.

natural in the context of hidden variables theories like Bohmian mechanics.

Historically, it is true that hidden variables theorists have thought of wave functions as physical fields. Most influentially, at the 1927 Solvay conference, De Broglie proposed what he called a pilot wave picture according to which the wave function is a wave spread out in space that more or less pushes particles along on their trajectories. This is a point of view that has been carried forward to the present day by some defenders of Bohmian mechanics.[39] But the more we think about the characteristics of realistic wave functions, the less, I would argue, this pilot wave picture seems plausible. So, let's pause for a moment to think about this.

So far, most of the wave functions I have used for illustration have involved representations of simple, i.e., single particle, systems. For the single particle case, as we've seen, wave functions are straightforwardly understood in terms of an assignment of values to points in ordinary space. Since Bohmian particles also take on locations in ordinary space, it is easy in these cases to make sense of a wave function as a pilot wave pushing a particle around. However, when we move to consider wave functions for multi-particle systems, the situation changes. Wave functions for complex systems, i.e., systems containing two or more particles, are not defined in terms of assignments of numbers to points in ordinary space. Rather, wave functions for complex systems (at least in nonrelativistic quantum mechanics) are defined in terms of an assignment of numbers to points in a different kind of space, a space with the structure of the configuration space of classical mechanics.

In classical mechanics, configuration spaces are used to produce simple representations of a system comprising many particles. Each point in configuration space corresponds to a listing of determinate positions for each particle in the system. So, using a configuration space representation, one may represent the locations of all

---

[39] See Bacciagaluppi and Valentini (2009).

particles in a system using a single point, and the trajectories of all of these particles over time using a single curve. Mathematically, each point in the configuration space is associated with a sequence of numbers $x_1, \ldots, x_n$, where relative to some coordinatization of the corresponding three-dimensional space, the first three numbers correspond to the x, y, and z coordinates for particle 1, the second three numbers correspond to the x, y, and z coordinates for particle 2, . . . , and the last three numbers correspond to the x, y, and z coordinates for the last particle. Thus the configuration space most commonly used in classical mechanics has dimension 3N, where N is the number of particles in the system under consideration. For a single-particle system, 3N = 3 and so we may think of the configuration space as ordinary space. But for multi-particle systems, 3N > 3, and so configuration space is distinct from ordinary space.

Just to be clear, although the dimensionality of the configuration space will generally be larger than three, it's not that the configuration space contains the three dimensions of our ordinary space plus some more. Configuration space is a different kind of space altogether. No three of the dimensions of a higher-dimensional configuration space correspond to any of the dimensions of our ordinary space. If we like, relative to some coordinatization of ordinary space and matching of configuration space dimensions with ordinary spatial dimensions, we might say that N of the 3N dimensions of configuration space correspond to the x dimension in ordinary space, another N of the 3N dimensions correspond to the y dimension, and the final N of the 3N dimensions correspond to the z dimension. But there is no absolute sense in which any single dimension of the 3N-dimensional configuration space corresponds to the x or the y or the z dimensions any better than any other. We will return to this in the chapters that follow.

Now, as I've said, wave functions are generally represented as fields spread out in a space with the structure of a classical configuration space. Let's illustrate this using a simple example. This

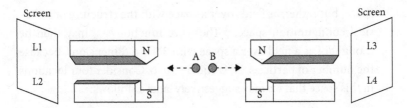

Figure 1.5. Two Spin Measurements

example will recur and give us much to think about throughout what follows.

Suppose we have two atoms whose spins we want to measure. Starting at a common origin, we send them in opposite directions toward two Stern-Gerlach detectors that will measure their spins along a given dimension (Figure 1.5).

Call our atoms "A" and "B" and the four locations that the atoms may be deflected to by the magnets L1, L2, L3, and L4. Suppose that after deflection, the wave function for the system is the following:

$$\psi_{AB} = \frac{1}{\sqrt{2}}|L1>_A|L4>_B + \frac{1}{\sqrt{2}}|L2>_A|L3>_B$$

The system is thus represented as being in a superposition of having atom A at L1 and atom B at L4, and atom A at L2 and atom B at L3. Note this is somewhat of an idealization as the state of a real system will be rather more spread out than this simple representation would indicate.

Consider a coordinatization of ordinary three-dimensional space such that:

L1: $(-1, 1, 0)$ L2: $(-1, -1, 0)$ L3: $(1, 1, 0)$ L4: $(1, -1, 0)$

Wave functions for multi-particle systems, we have noted, are not generally represented as fields over such locations in ordinary space, but rather, as fields over a space with the structure of a classical configuration space.[40] The wave function ψ_{AB} may then be thought of as a field over a space with 3N = 6 dimensions, because the number of particles, N, is equal to two. Consider four locations in this space that we may suggestively label as follows:

L13: (−1, 1, 0, 1, 1, 0)   L23: (−1, −1, 0, 1, 1, 0)
L14: (−1, 1, 0, 1, −1, 0)   L24: (−1, −1, 0, 1, −1, 0)

Here, we have assumed a correspondence such that the first three coordinates in the wave function's space correspond to the three (ordinary) spatial coordinates for atom A, and the second three coordinates in the wave function's space correspond to the three (ordinary) spatial coordinates for atom B. ψ_{AB} may then be understood as describing a field with amplitude $\frac{1}{\sqrt{2}}$ at two locations in this six-dimensional space: location L14 and location L23 (see Figure 1.6).

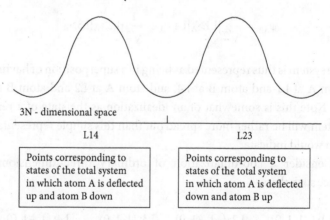

3N - dimensional space

L14   L23

| Points corresponding to states of the total system in which atom A is deflected up and atom B down | Points corresponding to states of the total system in which atom A is deflected down and atom B up |

Figure 1.6. Wave Function ψ_{AB}

[40] We will discuss why this is in the next chapter.

Of course, it is not easy to conjure up a visual representation of a wave in a six-dimensional space. But the key idea is that this wave will be peaked around two locations, one corresponding in the low-dimensional representation to a state in which atom A has been deflected to L1 and atom B to L4, and another corresponding in the low-dimensional representation to a state in which atom A has been deflected to L2 and atom B to L3.

In general, the ontology the wave function realist assigns to systems described by nonrelativistic quantum mechanics consists (at least) of a physical field (the quantum wave function ψ) defined not on the familiar three-dimensional space of our ordinary experience, but instead, on a different and higher-dimensional space. For nonrelativistic systems, this is a space with the structure of a classical configuration space with 3N dimensions (for some number N). When considering the state of the entire universe, this becomes the "universal wave function," typically denoted using the Greek capital letter Ψ. If it is the wave function of the universe that is fundamental, then since the number of total particles in the universe is generally estimated to be on the astonishingly large order of 10^{80}, the space this universal wave function Ψ inhabits, the space the wave function realist will argue is the fundamental physical space of our universe, is 3×10^{80}-dimensional.

Note that when the wave function realist approaches this 3N-dimensional space with ontological seriousness and considers it to be fundamental, she is no longer considering it anymore to be a literal space of configurations in the sense of classical mechanics. She *introduces* the space the wave function inhabits by way of talking about particle configurations in three-dimensional space, using the formula "the # of dimensions of the wave function's space = 3-times-N" as a heuristic to capture the dimensionality of the space.[41] But the dimensionality of the wave function's space

---

[41] As I keep saying, this is at least for the nonrelativistic case. We will consider alternative accounts of the dimensionality of the high-dimensional space of the wave function

is not determined by (nor grounded in) the number of particles in the universe. According to the wave function realist, it is not particles in three-dimensional space but the wave function in its space that is fundamental. And so the higher-dimensional space postulated by the wave function realist is not literally a space of configurations.

One key difference we may now see between wave function realists and their rivals consists in the former's rejection (at the level of basic ontology) of what Bell (1976) called "local beables." These are entities (*be*ables) that exist in or at some bounded three-dimensional spatial or four-dimensional spacetime region. The wave function, as the wave function realist understands it, is spread out over another kind of space, and so fails to be a local beable. According to the wave function realist, the wave function does not have any location in ordinary space or spacetime. The wave function realist believes that quantum theories provide good reason to revise our previous beliefs about fundamental ontology, and thus rejects local beables as ontologically basic entities.

Seeing all of this, let's now return to my claim that although there is a prima facie case for wave function realism as an interpretation of the ontological implications of Everettian and collapse approaches to quantum mechanics, the same may not be said for hidden variables approaches. Consider again Bohmian mechanics. In this theory, state representations have two basic parts: that giving the state of the wave function and that stating the positions of all of the particles. And there are two laws: the Schrödinger equation, which describes the evolution of the wave function, and the guidance equation, which describes how particle configurations evolve as a function of the state of the wave function. The Bohmian's motivation for adding facts about particle locations to the quantum formalism, recall, was to connect the formalism with facts about

realist, accounts more appropriate to capturing the full range of states in relativistic quantum theories, in Chapter 4.

our observations of the determinate locations of things in order to solve the measurement problem. From this it follows that these particles posited by the Bohmian are things in ordinary three-dimensional space. For the determinate values we find in our observations are determinate positions of things—scintillation patterns on a screen, pointers, tracks in a cloud chamber—in ordinary three-dimensional space.

So why does wave function realism fail to be a natural way to interpret the ontological implications of Bohmian mechanics? There are two reasons. First, to solve the measurement problem, Bohmians insist on a basic ontology of particles distributed at determinate locations in three-dimensional space. But, as we have just seen, wave function realists interpret the wave function as a physical field, a field in a different space altogether, a space of 3N dimensions with the structure of a classical configuration space. And this makes the relationship between the wave function and the particle locations, encapsulated in the Bohmian guidance equation, puzzling. How can a physical field in one space guide the behavior of particles in another space altogether? While it may be natural for the wave function realist to interpret the wave function as a pilot wave in the case of a single particle system, where the space of the wave function just is the space the particle inhabits, this model is not similarly natural for a more realistic multi-particle Bohmian mechanics. Here, the wave function is defined on and so would seem to inhabit a different space altogether.

Second, once one takes the step of interpreting the wave function as a physical field, one arguably has the resources to build up an ontology of particles with determinate locations.[42] So being a wave function realist itself would seem to undercut the motivation for being a Bohmian in the first place. One doesn't need to add particles with determinate locations to one's fundamental

---

[42] Arguably. This is the topic of Chapters 5–7.

ontology, since one can just see these particles as ontologically derivative entities built up out of the more fundamental wave function field.[43]

And so it shouldn't be surprising that Bohmians tend not to be wave function realists. This is not to say that Bohmians do not generally take a realistic attitude to the wave function, interpreting it as an objectively real phenomenon, one whose evolution over time is accurately described (for nonrelativistic systems) by the Schrödinger equation. But wave function realism, as we've seen, is a stronger position, one that says wave functions are real physical fields, spread out in a real physical space. It is more common for Bohmians to view wave functions as having some different metaphysical status. For example, Shelly Goldstein and Nino Zanghì have persuasively argued that Bohmians should take the wave function to be nomological, that is, to have a status more like that of a law of nature. It guides the particles not in the sense of being a physical field the particles may ride or be pushed along by, but rather it guides the particles in the sense of governing their behavior (Goldstein and Zanghì 2013). Once one adopts the view that the wave function is nomological or quasi-nomological, then it follows that the wave function is not the kind of thing that could

---

[43] This is what has come to be known in the literature as the "Everett in Denial" objection, after being presented by Harvey Brown and David Wallace (2005) in a paper criticizing Bohmian mechanics. As we will see in a moment, the objection doesn't really work against Bohmian mechanics as it is usually presented since Bohmians tend not to see the wave function as a physical field, i.e., as the kind of thing that could constitute material objects like particles. However, the objection does show a tension in combining Bohmian mechanics with wave function realism. Brown and Wallace accused Bohmians of being Everettians in denial because Bohmians accept all of the ontological and nomic structure as the Everettian, while mistakenly failing to accept that the wave function and its evolution are sufficient to capture all of the facts about particle positions. The Bohmian's hidden variables, according to Brown and Wallace, are thus just superfluous add-ons to an Everettian picture. As I note in the next paragraph, Brown and Wallace's conclusion doesn't follow if the wave function is not a physical field, but something more like a law. So, Bohmians that are wave function realists might be Everettians in denial, but most Bohmians are not. See Valentini (2010) for another response to Brown and Wallace (2005) from the pilot wave perspective. Valentini questions whether the wave function realist Everettian herself is able to view particles with determinate positions as ontologically derived from the wave function.

constitute three-dimensional particles or any other matter. And so particles must be added as additional elements of the Bohmian's ontology. They are not (as Brown and Wallace suggest) superfluous. We will discuss this view in more detail in later chapters.

Now just because there exists a more natural way of interpreting the kind of formalism Bohmians use in order to solve the measurement problem, this doesn't mean one cannot defend wave function realism as *a* way of interpreting Bohmian mechanics. Absolutely one can do so and indeed in his "Elementary Quantum Metaphysics," Albert motivated wave function realism especially as an account of the ontology of Bohmian mechanics. This is a way of providing an ontological interpretation of Bohmian mechanics that was also suggested by Bell. As the latter states in his "On the Impossible Pilot Wave":

> Note that in this theory [Bohm's 1952 theory] the wavefunction ψ has the role of a physically real field, as real here as Maxwell's fields. (1982, p. 162)

And in Bell's "Quantum Mechanics for Cosmologists":

> We then have a deterministic system in which everything is fixed by the initial values of the wave ψ and the particle configuration x. Note that in this compound dynamical system the wave is supposed to be just as "real" and "objective" as say the fields of classical Maxwell theory—although its action on the particles . . . is rather original. No one can understand this theory until he is willing to think of ψ as a real objective field rather than just a "probability amplitude." Even though it propagates not in 3-space but in 3n-space. (1981, p. 128)

Bell and Albert recommend taking the wave function seriously as a physical field in a high-dimensional space that evolves in accordance with the Schrödinger equation, and interpreting the guidance equation as describing how this wave function field determines the evolution of some additional determinate position values over time.

But what of the problem that the wave function seems to inhabit a difference space than the particles it is supposed to "guide"? The solution Albert (1996) proposes is that the Bohmian interpret the position values Q_k described by the guidance equation not to be the positions of a set of many particles in ordinary three-dimensional space, but rather the position values of something else inhabiting the same space on which the wave function is defined. The Q_k define the position degrees of freedom of a single particle, a so-called marvelous point. As we've seen, because the space on which the wave function is defined has the structure of a classical configuration space, it is possible to postulate a straightforward correspondence between the position of this single point in the wave function's 3N-dimensional space and the locations of a set of N particles in three-dimensional space, particles that may be said to constitute the things that record the results of measurements we make on quantum systems. What results is a wave function realist interpretation of Bohmian mechanics, an interpretation that assigns two basic entities to the ontology of quantum mechanics: the wave function, interpreted as a field on a high-dimensional physical space with the structure of a classical configuration space, and the marvelous point, guided around in this high-dimensional space under the influence of the wave function.

The marvelous point interpretation is an interesting take on Bohmian mechanics, but one that is certainly different from the picture one finds in those initially making the proposal to add facts about determinate positions directly into the quantum formalism. Nonetheless, it is a way the wave function realist may interpret Bohmian mechanics without raising questions about how a field in one space (the high-dimensional space of the wave function) may guide the behavior of a set of particles in another space altogether (ordinary three-space). Although it may still face the Everett in Denial objection, Albert's proposal shows it is in principle possible to apply wave function realism to any of the formulations of quantum mechanics that have been taken to solve the measurement problem in order to provide an ontological picture of a (nonrelativistic) quantum world.

1.6 A Higher-Dimensional Reality

The idea that wave functions are real, physical fields inhabiting spaces other than ordinary space shows wave function realism to be a startling and intriguing proposal. This ontological framework does not simply revise our classical picture of the kinds of entities out of which the world is made up, but also revises our view of the world's fundamental spatiotemporal structure. In what follows, we will see that this latter consequence of the wave function realist framework is one that motivates many of the objections against it. This is indeed the reason that Schrödinger himself turned away from taking the quantum wave function to represent a real physical field, shortly after developing the mathematical framework of wave mechanics. Schrödinger and other influential physicists of his day thought it simply implausible that the high-dimensional space required for wave functions to describe multi-particle systems could constitute a genuine, physical spatial framework. As Hendrik Lorentz put it in a letter to Schrödinger about his wave mechanics, comparing Schrödinger's wave mechanical approach to the rival matrix approach of Born and Heisenberg:

> If I had to choose now between your wave mechanics and the matrix mechanics, I would give the preference to the former, because of its greater intuitive clarity, so long as one only has to deal with the three coordinates x, y, and z. If, however, there are more degrees of freedom, then I cannot interpret the waves and vibrations physically, and I must therefore decide in favor of matrix mechanics. (Przibram 1934, p. 44)

Schrödinger ultimately agreed, saying, "I am long past the stage where I thought that one could consider the ψ-function as somehow a direct description of reality" (Fine 1996, p. 82, quoted in Monton 2013, p. 157).

In later chapters, we will consider more recent arguments against wave function realism to this effect. Critics claim that the wave function realist's postulation of a fundamental high-dimensional space makes the view empirically incoherent, i.e., unable to provide an account of the constitution of the devices we use to make the measurements that constitute the evidence for quantum theory. They argue that wave function realism wrongfully makes quantum theories about the nature and behavior of a field in a high-dimensional space rather than about the subatomic matter and radiation that constitute the world around us. In the chapters that follow, I will address both of these concerns. Wave function realism is not empirically incoherent. Nor does it miss the explanatory targets of quantum theories. If anything, the adoption of wave function realism permits us a way of seeing some fascinating new possibilities for the way unfamiliar and speculative kinds of physical objects may come to constitute the seemingly three-dimensional macroscopic objects around us, including ourselves. Moreover, rather than being an obstacle to providing a sensible interpretation of quantum theories, it is this very aspect of wave function realism, that it makes use of a higher-dimensional framework, that allows the wave function realist to provide an account of the world that has certain appealing metaphysical features, dissolving some of the most foundational puzzles about quantum theories and their relation to the world we observe and take ourselves to know.

2

The Argument from Entanglement

2.1 Entanglement as the Characteristic Feature of Quantum Theories

In this chapter, we will move beyond the prima facie case for wave function realism, which stems from noting the ubiquity and success of wave function representations in formulations of quantum mechanics that solve the measurement problem. Here we will try to be less tentative and instead look at what might be thought of as an all things considered case for this approach to providing an ontological interpretation of quantum theories. One argument that some proponents of wave function realism have given is that if one wants a realist interpretation of quantum mechanics, then the phenomenon of quantum entanglement forces wave function realism on one. That is, it is not possible to give an objective and complete description of the full range of quantum states, which include entangled states, without recognizing the reality of the wave function (here, the universal wave function) and taking it seriously as a physical field in a high-dimensional space. I will call this the argument from entanglement.

In earlier work, I myself gave an argument like this (Ney 2012) and it is easy to find elsewhere in the literature. Jill North (2013) is especially explicit. However, I now think that the case for wave function realism is not as straightforward as this argument would make it seem. And so in Chapter 3, I will present what I instead favor as a more promising way of providing an all things considered case for wave function realism, namely, by arguing that wave function realism presents a picture of the quantum world that

The World in the Wave Function. Alyssa Ney, Oxford University Press (2021). © Oxford University Press.
DOI: 10.1093/oso/9780190097714.003.0002

has some intuitively nice metaphysical features, features that the main rival approaches to interpreting quantum theories lack: separability and locality. But first, I will consider the argument from entanglement. And so to see how this goes, we will first need a bit of technical background about the concept of entangled states.

Schrödinger famously called entanglement the "characteristic" feature of quantum theories (1935). One common way of understanding entangled states takes them to be states of two or more systems where, due to some interaction in their past, there now exists a correlation between the values they are predicted to take on upon measurement. Entangled states may be represented as superpositions of states of multi-particle systems.

For illustration, we may build on the case of the two atoms we considered in Chapter 1. This is a case Bohm used in his 1951 textbook *Quantum Theory* to illustrate the core idea behind the famous Einstein-Podolsky-Rosen thought experiment of 1935. Here is Bohm:

> Suppose that we have a molecule containing two atoms in a state in which the total spin is zero and that the spin of each atom is $\hbar/2$. Roughly speaking, this means that the spin of each particle points in a direction exactly opposite to that of the other, insofar as the spin may be said to have any definite direction at all. Now suppose that the molecule is disintegrated by some process that does not change the total angular momentum. The two atoms will begin to separate and will soon cease to interact appreciably. . . . When the atoms separated, each atom would continue to have every component of its spin angular momentum opposite to that of the other. The two spin-angular-momentum vectors would therefore be correlated. . . . Suppose now that one measures the spin angular momentum of any one of the particles, say No. 1. Because of the existence of correlations, one can immediately conclude that the angular-momentum vector of the other particle (No. 2) is equal and opposite to that of No. 1. (1951, p. 614)

In this scenario, our atoms are in an entangled state, the singlet state, in which the two are entangled with respect to their spin along some particular axis, say the x-axis. Atoms in such a state may be represented by the following wave function:[1]

$$\psi_{singlet} = \frac{1}{\sqrt{2}} |x - up\rangle_A |x - down\rangle_B - \frac{1}{\sqrt{2}} |x - down\rangle_A |x - up\rangle_B$$

The Born rule will then tell us that were we to measure the x-spin of these atoms, we would have a 50% chance of finding the first x-spin-up and the second x-spin down, and a 50% chance of finding the first x-spin-down and the second x-spin up.

As discussed in the previous chapter, we may also consider these atoms at some point after they are sent through the Stern-Gerlach devices that will measure their spin. Using the same location labels as before, we may take their resulting position state to be:

$$\psi_{AB} = \frac{1}{\sqrt{2}} |L1\rangle_A |L4\rangle_B + \frac{1}{\sqrt{2}} |L2\rangle_A |L3\rangle_B$$

By the Born rule, we have a 50% chance of finding the first at L1 and the second at L4, and a 50% chance of finding the first at L2 and the second at L3. These atoms that began in an entangled spin state are now entangled with respect to their positions.

In quantum mechanics, it is important to emphasize, entangled states are not exotic or rare. Rather, because of the frequent interactions between physical systems, entanglement is the norm and so the metaphysical consequences of entanglement are very much a central issue for anyone wishing to provide an ontological interpretation of the various formulations of quantum mechanics.

---

[1] I will continue, as in the previous chapter, to label the atoms "A" and "B."

2.2　The Necessity of Wave Function Realism?

As I mentioned, some have argued that it is essential, in order to interpret quantum systems realistically and objectively, that we take the wave function realist approach, that is, that we interpret wave functions as real physical fields, and thus see such wave functions as at least part of the fundamental stuff out of which other material objects are constituted. As Albert puts it:

> It has been essential . . . to the project of quantum-mechanical *realism* (in *whatever* particular form it takes—Bohm's theory, or modal theories, or Everettish theories, or theories of spontaneous localization), to learn to think of wave functions as physical objects *in and of themselves*. (1996, p. 277)

But why is it essential, when one regards quantum mechanics realistically, to take on wave function realism? One argument that has been given derives from the consideration of entanglement. The argument is that if one is to take an objective or realist attitude toward entangled states, then one must regard wave functions as real physical fields in a high-dimensional space that make up the stuff in our universe. There is no other adequate way to make sense of the full range of pure quantum states, which include entangled states, ontologically, as objective states of matter.[2]

Here is a way to see this. Suppose one wanted to hang onto an objective three-dimensional interpretation of a system in an entangled state. For illustration, return to Bohm's case of the two atoms that we have described as being in an entangled state of position:

---

[2] Pure quantum states are to be contrasted with mixed states. The latter are statistical mixtures of pure quantum states, those in which a system is described as having a certain probability of being in one or another pure state.

$$\psi_{AB} = \frac{1}{\sqrt{2}} |L1>_A |L4>_B + \frac{1}{\sqrt{2}} |L2>_A |L3>_B$$

One option is to say that in this case there are objective and definite facts about the locations of the two particles A and B. Either
A is at L1 and B is at L4 or A is at L2 and B is at L3. The entangled
state ψ_{AB} merely records our uncertainty about A and B's locations
and describes nothing objective that is wavelike. This would not be
to take a realistic approach to facts about quantum entanglement.
This would instead be to take a subjectivist or anti-realist or epistemic approach to quantum entanglement and the wave function.
As mentioned in Chapter 1, this approach seems undermined by
experimental results proving the objective existence of wave-like
quantum phenomena. I will say a bit more about it shortly, but for
now, what we want to consider is what one should say if one wants
to use quantum mechanics to inform one's ontology and interpret
quantum entanglement as a real, objective phenomenon, as opposed to one merely reflecting the limits of our knowledge.

Adopting a realist attitude, one could instead say that the
entangled state ψ_{AB} objectively describes a situation in which these
atoms fail to have any definite locations. Recalling Schrödinger's
comment about clouds and fog banks, a natural first step would be
to explain this in terms of the atoms each being rather smeared out
in space. A is smeared out so as to be partly at L1 and partly at L2.
B is smeared out so as to be partly at L3 and partly at L4. In each
case, we take the coefficients $\frac{1}{\sqrt{2}}$ to suggest that each atom has half
of its existence or being at each of its respective locations. Call this
interpretation of entangled systems the *simple field interpretation*.

To see why the simple field interpretation is inadequate, we
need to contrast this example with another in which there are a
pair of atoms that have been created in some measurably distinct
spin state, and yet for which the simple field interpretation would

yield the same ontology. For example, consider the following wave function:

$$\psi_{A'B'} = \frac{1}{2}|L1>_{A'}|L3>_{B'} + \frac{1}{2}|L1>_{A'}|L4>_{B'}$$
$$+ \frac{1}{2}|L2>_{A'}|L3>_{B'} + \frac{1}{2}|L2>_{A'}|L4>_{B'}$$

By the Born rule, an experimenter measuring the atoms in this state should expect to see any of the four position combinations with probability ¼ each. Clearly then, ψ_{AB} and $\psi_{A'B'}$ are different quantum states. They are measurably different, since they yield different probabilities for measurement results. And thus, if we are interpreting quantum states objectively, then we should say that they correspond to different facts about what there is.

Yet the simple field interpretation conflates these two states. In ψ_{AB} just as in $\psi_{A'B'}$, the first atom is certain to be found at either L1 or L2 with equal probability, and the second atom is certain be found at either L3 or L4 with equal probability. And so, the simple field interpretation takes the ontology of either state to consist of one atom A/A' smeared out equally over L1 and L4, and a second atom B/B' smeared out equally over L2 and L3. What the simple field interpretation misses is the fact that in the entangled state, ψ_{AB}, the positions of the two atoms are correlated. In ψ_{AB}, it is not just that A has a 50% chance of being found at L1, but that the whole system has a 50% chance of having one atom at L1 and one atom at L4. Only in $\psi_{A'B'}$ do the atoms have any chance of being found at the pair of locations L1 and L3.

Wave function realism, on the other hand, yields distinct ontologies for the quantum states ψ_{AB} and $\psi_{A'B'}$. This is accomplished by the fact that the wave function realist understands quantum systems not as multiple fields spread out over three-dimensional space, but rather each as a single field spread out over a high-dimensional space with the structure of a classical configuration space. In the wave function's 3N-dimensional space, we consider there to be four locations which, in Section 1.5, we suggestively labeled L13, L14, L23, and L24. In ψ_{AB}, the wave function is spread evenly over two locations in the 3N-dimensional space, L14 and L23. In $\psi_{A'B'}$, the wave function is spread evenly over all four locations L13, L14, L23, and L24. So for the wave function realist, these measurably distinct quantum states correspond to different high-dimensional ontologies.

The trick to finding a field interpretation that does not conflate empirically distinct quantum states then is to view quantum systems ultimately as single fields spread out in a higher-dimensional space. North uses this point to motivate the following argument:

> In quantum mechanics . . . we must formulate the dynamics on a high-dimensional space. This is because quantum mechanical systems can be in entangled states, for which the wave function is nonseparable. Such a wave function cannot be broken down into individual three-dimensional wave functions, corresponding to what we think of as particles in three-dimensional space. That would leave out information about correlations among different parts of the system, correlations that have experimentally observed effects. Only the entire wave function, defined over the entire high-dimensional space, contains all the information that factors in the future evolution of quantum mechanical systems.
>
> Following the principle to infer, at the fundamental level of the world, just that structure and ontology that is presupposed by the dynamics, we are led to conclude that the fundamental space of a world governed by this dynamics is the high-dimensional one. The fundamental ontology, which includes the wave function, then lives in it. (2013, pp. 190–191)

And similarly, I concluded in previous work:

> [Entangled] states can only be distinguished, and hence completely characterized in a higher-than-3-dimensional configuration space. They are states of something that can only be adequately

characterized as inhabiting this higher-dimensional space. This is the quantum wave function. (2012, p. 556)

But is the wave function as a physically real field in higher dimensions required by quantum representations? Can entangled states "only be distinguished" in such a higher-dimensional ontology?

The answer to these questions is: absolutely not. There are rival ontological approaches to wave function realism, rival approaches that are capable also of capturing and distinguishing the differences between measurably distinct quantum states. The simple field interpretation is a natural place to start in capturing the ontology of quantum theories, but it is not wave function realism's only or even most plausible interpretational rival. In Chapter 3, I will present an argument to the effect that wave function realism is to be preferred over these other approaches. In the present chapter, I will present a positive survey of what the most promising rivals to wave function realism look like and how they are motivated. The very existence and empirical adequacy of these alternatives undermines the argument from entanglement, demonstrating that it is not true that wave function realism is essential for providing a realist characterization of quantum mechanics, including entangled states.

2.3 Rivals: The Primitive Ontology Approach

The primitive ontology approach of Detlef Dürr, Shelly Goldstein, Nino Zanghì, Valia Allori, and Roderich Tumulka (Dürr et al. 1992, Allori et al. 2008, and Goldstein and Zanghì 2013) starts from the claim that the ontology of quantum theories consists primarily of entities in ordinary space or spacetime. In other words, it consists of local beables. Advocates of the primitive ontology view thus emphasize an element of continuity between the ontology of classical physics and that of quantum physics.

These local beables constitute what Dürr et al. call the "primitive ontology" of a quantum theory: they are what a given theory is primarily about, what it is aimed at explaining the behavior of, and what constitutes material objects according to that theory. This approach to interpretation was originally developed in Dürr, Goldstein, and Zanghì 1992, as part of a defense of Bohmian mechanics. They argued that in order for quantum theories to connect to the objects these theories were initially aimed at describing, they must include a representation of local beables. Dürr, Goldstein, and Zanghì thus argued that orthodox quantum mechanics and other collapse approaches, as well as Everettian quantum mechanics, as these stood, were defective in only describing the states and behavior of the wave function and failing to include representations of a primitive ontology and its evolution. By contrast, Bohmian mechanics explicitly includes a primitive ontology, with its description of the determinate positions of particles and their evolution over time in the guidance equation.

The wave function is real in Bohmian mechanics, but it is not primitive ontology. It is not the primary thing the theory is about, nor what constitutes material objects. Instead the wave function is included in the ontology of Bohmian mechanics, according to Dürr, Goldstein, and Zanghì, solely to explain how the primitive ontology evolves:

> According to (pre-quantum-mechanical) scientific precedent, when new mathematically abstract theoretical entities are introduced into a theory, the physical significance of these entities, their meaning insofar as physics is concerned, arises from their dynamical role, from the role they play in (governing) the evolution of the more primitive—more familiar and less abstract—entities or dynamical variables. For example, in classical electrodynamics the *meaning* of the electromagnetic field derives solely from the Lorentz force equation, i.e., from the field's role in governing the evolution of the positions of charged

particles. . . . That this should be so is rather obvious: Why would these abstractions be introduced in the first place, if not for their relevance to the behavior of *something else*, which somehow already has physical significance?

Indeed, it should perhaps be thought astonishing that the wave function was not also introduced in this way; insofar as it is a field on configuration space rather than on physical space, the wave function is an abstraction of even higher order than the electromagnetic field.

But, in fact, it was!

What we regard as the obvious choice of primitive ontology— the basic kinds of entities that are to be the building blocks of everything else [footnote: Except, of course, the wave function.]—should by now be clear: Particles, described by their positions in space, changing with time—some of which, owing to the dynamical laws governing their evolution, perhaps combine to form the familiar macroscopic objects of daily experience. (1992, pp. 7–9)

Tim Maudlin has advocated a view similar to the primitive ontology approach, although he refers to "primary" rather than "primitive" ontology (2013, pp. 143–147). Again the idea is that quantum mechanics must include some basic ontology of matter in space or spacetime:

In the standard versions of [Bohmian mechanics], there is a local ontology (particles, say) in a low-dimensional space-time, organized to form pointers on boxes, patterns of ink on paper, and so on. The quantum state [i.e. what is represented by the wave function] is introduced into the theory as part of the physical account of how the local beables come to be arranged as they are. . . . That is, the quantum state is introduced as part of Secondary Ontology while proposing a physical account of why the Primary Ontology behaves as it does. Even more precisely,

the quantum state is introduced as a part of a physical account
of why the configuration of local beables evolves as it does.
(2013, p. 147)

So those who take the primitive (or primary) ontology approach
to answering the ontological questions start from a postulation of
local beables. The wave function is interpreted as real, as objective,
and thus as genuine ontology.[3] But it is not everything. It cannot be
the sole ontology of the theory, since physical theories must include
some local beables. That is what they are primarily about.

Although the primitive ontology approach was developed in the
context of a defense of Bohmian mechanics, later work by Goldstein,
Zanghì, and their collaborators, Allori and Tumulka, extended the
approach to other strategies for solving the measurement problem,
particularly the Ghirardi-Rimini-Weber (GRW) spontaneous collapse approach and Everettian or many worlds theories. Since the
standard versions of such theories only include representations and
laws describing the behavior of the wave function, these authors
proposed the theories be modified to include facts about local
beables as well as laws similar to the Bohmian guidance equation
describing how these beables evolve as a function of the state of the
wave function. In the case of the GRW theory, they call what they
deem an unsatisfactory version of the theory without a primitive
ontology GRW_0, and propose a couple of alternative versions of
GRW with primitive ontologies (Allori et al. 2008).

For example, in GRW_m, one supplements the theory with
representations of a matter density field spread out over space. This
is a field that can serve to constitute material objects, one that tends

---

[3] In metaphysics, sometimes one distinguishes between ontology (narrowly construed
as concerning which entities or objects exist) and other elements or aspects of objective reality. All of what is objective and real is what Sider (2011) calls "structure." Here,
I follow those making the primitive/primary versus non-primitive/secondary ontology
distinction in using "ontology" in a broader sense to include whatever is real and objective, or in Sider's sense, structure. This may include nomological structure.

to spread out over time, but bunches up around definite locations in space as the wave function undergoes hits. GRW_m includes a law describing the matter density field's evolution as determined by the nonlinear GRW evolution of the wave function. This suggestion was first proposed by Ghirardi and his collaborators, Benatti et al. in 1995.

Alternatively, Bell (1987) proposed that the GRW theory be supplemented with explicit representation of spacetime events corresponding to the spontaneous hits or collapses of the wave function. Bell calls these spacetime events "flashes." Thus, GRW_f is a version of the GRW theory with explicit representation of such flash events and a law describing when and where their occurrence is likely to take place according to the evolution of the wave function.

For an Everett-style relative state or many worlds approach to solving the measurement problem, Goldstein, Zanghì, and their collaborators propose we again supplement the wave function representation and Schrödinger equation with a description of the state and evolution of a matter density field that is guided by the wave function. In this theory, the matter field evolves linearly and deterministically so as to constitute multiple worlds corresponding to distinct possibilities for measurement (Allori et al. 2011). Thus, the primitive ontology for the theory they call S_m (where "S" stands for Schrödinger equation) will be similar to that for GRW_m except here the matter field will evolve deterministically, not stochastically.

Since those defending the primitive ontology approach propose not only new answers to the ontological questions, but also new quantum formalisms, we should modify our list of theories from Table 1.1 accordingly (see Table 2.1).

Thus, according to the primitive ontology approach, local beables of one sort or another constitute the primitive ontology of any physical theory, and so the quantum formalism should include laws describing the evolution of these local beables. This primitive

Table 2.1 Versions of Quantum Mechanics

Approach to Solving the Measurement Problem	State Representation	Law(s) of State Evolution
Many-worlds quantum mechanics:		
I (S_0)	Wave function	Schrödinger equation
II (S_m)*	Wave function + representation of matter density field	Schrödinger equation Matter guidance equation
Spontaneous collapse theory (GRW):		
I (GRW_0)	Wave function	Stochastic GRW equation
II (GRW_m)*	Wave function + representation of matter density field	Stochastic GRW equation Matter guidance equation
III (GRW_f)*	Wave function + representation of flashes	Stochastic GRW equation Flash equation
Bohmian mechanics*	Wave function + particle configuration	Schrödinger equation Particle guidance equation

Notes: A similar table appeared in Ney and Phillips (2013), p. 459.

Those formalisms preferred by Goldstein, Zanghì, and their collaborators are marked with an asterisk (*).

ontology will vary according to the strategy one pursues for solving the measurement problem: particles for nonrelativistic Bohmian mechanics, matter density fields or flashes for collapse theories, and matter density fields for many worlds approaches.

While wave functions are not primitive ontology, there are objective facts about which wave functions describe physical systems and how they guide the behavior of matter. This is important, and it is what allows the primitive ontology theorist a way to capture the distinction between measurably distinct states of the kind described in Section 2.2. Wave functions are real, and those for entangled systems guide or constrain the primitive ontology to evolve differently than wave functions for nonentangled systems. This explains the correlations we find in measurements on these systems.

It is for this reason that those adopting the primitive ontology framework (and some of the other approaches I describe below) sometimes complain about the use of the name 'wave function realism' to apply solely to views according to which the wave function is a physical field on a high-dimensional space, noting they too are realists about the wave function, taking it to be a real, mind-independent element of quantum ontology. In a sense, this complaint is fair, but by now the terminology has become so entrenched that I will continue to use it. And anyway, in defense of this terminology, one thing that distinguishes the status of the wave function on the wave function realist view from how it is viewed in primitive ontology or other realist approaches to interpretation is that for the wave function realist, the wave function is real in the classical sense of being *res*- or thing-like; it is a substance, *re*-al. This contrasts with its status on other interpretational approaches in which it occupies one or another distinct ontological category; rather than being *res*, it viewed as law, property, or a pattern of relations.

To many, the primitive ontology framework has appeal over wave function realism in providing pictures of the world according to quantum theories that are intuitive in certain respects. According to most applications of the approach, the fundamental entities of the theory inhabit our familiar space or spacetime, and the macroscopic objects we observe may be built out of these basic constituents (particles or matter fields) in straightforward ways. This claim will be evaluated in Chapter 5.

2.4 Rivals: Holisms

A distinct class of approaches to the interpretation of quantum theories sees a fundamental holism as the ontological lesson one should take from quantum entanglement. Return again to Bohm's example of the atoms A and B in an entangled state of spin and then position. What we know about such situations is that unless a suitable kind of interaction intervenes to break the entanglement between A and B, this connection will be permanent and persist no matter how far away A and B drift from each other. If A is measured spin-up along a given axis, B will be measured spin-down and vice versa. There is a connection that persists between these objects that has suggested to some that in a significant sense, they are not really distinct objects at all.

Now we know that there is nothing particularly special about cases in which objects may continue to bear relations to each other even upon spatial separation. My sister, my brother, and I have certain deep similarities, both physical and psychological, that will persist no matter how far apart we are separated and how long we live.

But cases described by quantum entanglement are different. When quantum systems are entangled, there are relational facts that do not obtain in virtue of any properties the individuals bear intrinsically, or on their own. That my sister, brother, and I all root for the same baseball team is not a brute relational fact about us but obtains in virtue of each of our individual preferences and behaviors. But in the case of Bohm's entangled atoms, there is nothing about the way A or B is individually on its own; there are no individual facts about these atoms' positions or spins, that fix the relational facts. The relational facts about the correlations in both the spin and position of A and B appear to be brute facts about the pair.[4]

---

[4] As we've seen, the wave function realist denies this fact is brute. She argues it obtains in virtue of the features of a field, the wave function, in a higher-dimensional space. In

This has suggested to some that according to quantum theories, entangled wholes such as the pair of atoms in Bohm's case, collections of particles caught up in measurement processes, perhaps even the universe as a whole, bear properties that are more fundamental than the systems out of which they are composed. For them, it is the *pair* of entangled atoms in Bohm's case that bears a definite and fundamental spin feature, that of having opposite x-spin. And there is nothing more fundamental to be said about the individual spins of the atoms on their own.

Brad Monton has used this point to argue that we should interpret the wave function in this way as corresponding to a property, one had by whole collections of particles in entangled states, a property not determined by any more fundamental properties of individual particles or atoms. He says:

> That's how the wave function fits into my picture: the wave function doesn't exist on its own, but it corresponds to a property possessed by the system of all the particles in the universe (or whatever closed system you're interested in). (2006, p. 779)

It is the fact that different closed systems instantiate different such properties that allows Monton to capture distinctions between entangled and nonentangled states.

Mauricio Suárez too, has recommended such a position, that we should think of the wave function not as a describing properties of individual systems (as the simple field interpretation had it) but instead complex properties of whole systems. In his "Bohmian Dispositions," an interpretation of quantum theories tailored to Bohmian mechanics, Suárez argues that we should construe the wave function as picking out a dispositional property of composite systems:

this section, we take a step back from wave function realism and consider what someone might say about entanglement without moving to a higher-dimensional ontology.

On this view, the wavefunction is not a law, and has no nomological force. It has merely a descriptive, or representational,
function concerning the state of the physical particles in 3-d
space. Yet, it does not represent any distinct object per se in 3-
d space—neither a field nor a wave nor even the particle itself.
And it certainly does not represent the state of an ultra-particle in
N-dimensional configuration space. Its function is rather to represent . . . the properties of the 3-d particles, including crucially
a series of dispositional properties over and above the particles'
positions. (2015, p. 3215)

This includes facts about the velocities that particles would have at
later times, had their initial positions been different, facts that can
be calculated using the guidance equation of Bohmian mechanics.

Some have gone farther than Monton and Suárez, arguing that entanglement recommends an even more revisionary metaphysics. An
old view about the nature of substance tracing to Aristotle claimed
that substances are the things that are the fundamental bearers of
properties (Heil 2012). If this is correct, then Monton and Suárez's
points would seem to lead to a variety of holism. Since in cases of
quantum entanglement, it is entangled pairs rather than individuals
that are the bearers of definite properties, it is the wholes not the
individuals that we should regard as the fundamental substances.

Jonathan Schaffer, individually and in collaborative work with
Jenann Ismael (2016), has argued that in this way monism is the
ontology suggested by quantum theories. In other words, it is not
merely wholes that are fundamental rather than the individuals
that seem to compose them, but it is rather the entire universe itself that is the fundamental substance. In Schaffer's "Monism: The
Priority of the Whole," he uses the following argument to establish
what he calls priority monism:

The cosmos is in an entangled state.
Entangled systems are fundamental wholes.

Therefore,

The cosmos is a fundamental whole. (2010, pp. 53–55)

That the universe or cosmos as a whole is in an entangled state is claimed to follow from the supposition that the initial state of the universe, the Big Bang, was one in which everything interacted and the linear evolution of the Schrödinger equation preserves this initial entanglement.[5] Note that the monism being advocated here is one that Schaffer refers to as a "priority monism" rather than an "existence monism." In other words, Schaffer is not intending to argue that it is only the universe or cosmos as a whole that is real. He accepts the existence of subsystems, be they microscopic atoms or particles or macroscopic organisms and measuring devices, and that these subsystems may also participate in entangled states. His claim is only that these subsystems are less fundamental than the cosmos as a whole, that they are constituted out of or, perhaps better, exist merely as abstractions from features of the cosmic whole. Schaffer thus endorses a layered ontology, with existence stratifying into divisions according to what is fundamental and what exists derivatively in virtue of the fundamental.[6]

Don Howard has also invited us to regard the phenomenon of quantum entanglement as suggesting a holist or monistic position. However his argument for holism is rather different from Schaffer's. We will discuss Howard's argument, which stems from a desire to avoid action across spatial distances or nonlocality in Chapter 3. Here I will only mention one key difference between the quantum holism adopted by Howard and that proposed by Schaffer. Unlike Schaffer, what Howard needs for his argument to work is what Schaffer would call existence holism, that it isn't that quantum wholes are more fundamental than their parts, but that quantum

---

[5] Claudio Calosi (forthcoming) argues that this is compelling only if we assume certain solutions to the measurement problem. See also my (2010) for further discussion of Schaffer's argument.

[6] See Schaffer (2009).

entanglement shows that it is only these "wholes" that exist. In a sense it may seem that this isn't even a version of holism. Since the parts aren't real, this means that there aren't wholes either, but only simple objects that are mistakenly viewed by us as wholes.

Of course, the wave function realist also advocates an interpretation of quantum theories according to which what they are telling us is that fundamentally, there is only one thing, the quantum wave function. And so in a sense wave function realism too is a version of monism or holism.[7] But the kind of view adopted by the wave function realist is in a significant sense more radical than that adopted by the monists or holists considered in this section, as the latter regard the quantum wholes or cosmos as existing in the low-dimensional space or spacetime of our ordinary experience.

2.5 Rivals: Relational Approaches

In the 1980s, Paul Teller offered a distinct approach to providing an ontological interpretation of quantum theories that would support facts of entanglement, one that he also classified as a variety of holism, a so-called relational holism. Teller complained that earlier versions of holism were problematic:

> Holism has always seemed incoherent, for it seems to say that two distinct things can somehow be entangled or intermeshed so that they are not two distinct things after all. Yet apparent unintelligibility does not prevent holism from recurring, not only in the

[7] Two caveats: (1) For it to be monistic assumes a non-hidden-variables approach to solving the measurement problem. The marvelous point interpretation of Bohmian mechanics, as we have seen, has it that there are two fundamental objects, the wave function and the marvelous point. (2) Whether wave function realism is properly regarded as a version of holism depends on whether the wave function realist regards the relationship between the wave function and less fundamental objects (e.g., particles and atoms) as standing in whole-part (i.e., mereological) relations. I will argue in Chapter 7 that she should.

work of philosophers of East and West, but also in what quantum mechanics seems to many of us to be saying about the world. (1986, p. 73)

Relational holism is supposed to provide a version of holism that lacks this seeming incoherence. The idea is to allow that entangled pairs are fundamentally distinct entities, but draw from the facts of entanglement the conclusion that quantum systems possess irreducible relations that do not hold in virtue of any intrinsic features of the individuals constituting those systems. It is a brute fact that Bohm's entangled atoms have opposite spins. But we needn't infer from this fact that the atoms are mere abstractions from some more fundamental oneness.

Teller defines relational holism as the view that "collections of objects have physical relations which do not supervene on the non-relational physical properties of the parts" (1986, p. 73), arguing that cases of quantum entanglement provide genuine cases in which collections of objects have what he calls "inherent relations," relations whose instantiation are not determined by intrinsic features of their relata.[8] Different entangled (or nonentangled) states are captured by different patterns of inherent relations. We may note that since holism is traditionally the view that wholes are, in some sense, more fundamental than their parts, what Teller is proposing isn't strictly speaking a version of holism. It is for this reason that I classify Teller's view as one among several "relational" approaches to drawing ontological lessons from the phenomenon of quantum entanglement.

More recently, the view that quantum entanglement should be interpreted as indicating facts about fundamental relations that

---

[8] In recent decades, philosophers have recognized the inadequacy of the notion of supervenience for capturing ontological theses of this kind (see, e.g., Fine 2001, Schaffer 2009, and Sider 2011). For this reason, one might instead characterize relational holism as the view that collections of objects have physical relations which fail to obtain in virtue of the nonrelational physical properties of the parts.

obtain brutely and not in virtue of individual features of their relata has become quite popular. One widely defended position in the philosophy of science is ontic structural realism (Ladyman 1998, Esfeld 2004, Ladyman and Ross 2007, French 2014, McKenzie 2017).[9] Ontic structural realism is a very general framework for interpreting physical theories, one that is not motivated solely by its utility in helping us understand the ontological consequences of quantum theories, although advocates have certainly appealed to quantum phenomena as supporting their position.

The master argument for structural realism promotes it as the best way of being a scientific realist in general (Worrall 1989). Scientific realists are those who believe that our best scientific theories are true (or approximately true) and that the entities they describe exist in at least roughly the ways these theories say they do. They see our best scientific theories as having remarkable success, as they have made what are often quite risky predictions about the results of future experiments and observations, and these predictions have later met with confirmation. Scientific realists see such inductive successes as support for a realist attitude toward a given theory, for how could these theories have had such successes if they weren't tracking something true (Putnam 1975)? And yet, scientific realists at the same time must face the fact that over the history of science, many theories have met such successes and later been proven largely wrong. The history of science is littered with posits that have since been rejected: caloric, phlogiston, the luminiferous ether. This latter point has seemed to some to license a more pessimistic attitude, that we should be skeptical that our current theories are tracking anything true about the world (Laudan 1981). Worrall noted, however, that although these historical upheavals give us reason to be skeptical that our best scientific theories give us an accurate representation of the kinds of things

---

[9] Ladyman explicitly makes the connection with Teller's work in his article "What Is Structural Realism?" (1998).

that exist, there is a good deal that tends to be retained through theory change. This, he argues, is what can explain the predictive successes of past theories. What gets retained as science evolves, Worrall argues, are not facts about the kinds of entities that exist, but instead structural relationships, including those we may sometimes regard as laws of nature.

On the structural realist view, what physicists really discover are the relationships between phenomena expressed in the mathematical equations of their theories (1989, p. 122). And so, Worrall reaches the conclusion that the kind of scientific realism most worth adopting is one that takes only the structural claims made by our best scientific theories to be (approximately) true.[10] Successful theories teach us about relational structures. We should remain agnostic about the kinds of entities instantiating these structures.

In his article, "What Is Structural Realism?" James Ladyman advocates transforming this structural realism into an ontological position. It is not merely that our best scientific theories can only teach us about structure and so we must be agnostic about other parts of their descriptions of the world. Rather, what these theories are telling us is that fundamentally, there is only structure:

> We need to recognize the failure of our best theories to determine even the most fundamental ontological characteristic of the purported entities they feature. . . . This means taking structure to be primitive and ontologically subsistent. (Ladyman 1998, pp. 419–420)

Here, Ladyman has quantum theories in mind, including those quantum field theories making up the Standard Model of particle physics, as examples of the "best theories" in science. He notes that this position resembles not only the proposal of Teller but also one of Howard Stein, who once commented:

[10] Note that this is a different sense of "structure" than that attributed to Sider above.

But if one examines carefully how phenomena are "represented" by the quantum theory . . . then . . . interpretation in terms of "entities" and "attributes" can be seen to be highly dubious. . . . I think the live problems concern the relation of the Forms . . . to phenomena, rather than the relation of (putative) attributes to (putative) entities. (1989, p. 59)

So, structural realism, as a general framework in philosophy of science, gets motivated as an ontological position, ontic structural realism, in the context of quantum theories.

Ladyman and Steven French as well have argued that rather than pursuing wave function realism, it is better to take a structuralist approach to interpreting the ontology of quantum theories. For if one takes the wave function as real or substance-like, then one has to address questions about the kind of substance it is, what kind of space it inhabits, should we be wave function realists or monists, and that the answers to such questions just aren't to be found in physics. These are questions that get bypassed altogether when one takes on a structuralist perspective (see Ladyman 1998, p. 419). French fleshes out his structuralism in the following way:

In general, this structure is constituted by the laws and associated symmetry principles of our fundamental theories. . . . In particular, in this case, it will encompass Schrödinger's equation. From the structuralist perspective, this should not be seen as governing the evolution of the wave function *qua* object; rather, it expresses the dynamic nature of the structure itself. . . . [F]or many realists, laws are a guide to the fundamental nature of a world, but for the structuralist, laws *are* the fundamental nature of the world. (2013, p. 87)

French's laws-and-symmetry-based approach to structuralism is worked out in more detail in his 2014 book *The Structure of the World* and is eliminativist about objects. However, other ontic structural realists pursuing interpretations of quantum theories have (like Teller) retained fundamental objects, arguing that these instantiate only relations, never intrinsic properties (e.g., Esfeld 2004). Either way, the facts about quantum entanglement are taken to be genuine, but they are captured metaphysically by structural features, not by the existence of some single object, the wave function, that is fundamental or real, nor by some collection of objects possessing intrinsic features.

2.6 Rivals: Spacetime State Realism

Another interpretive framework adequate to distinguishing entangled states, one that will play a large dialectical role in later chapters, is the spacetime state realism advocated by David Wallace and Chris Timpson (2010) and Wayne Myrvold (2015). On this view, the wave function is understood not as a field on a high-dimensional space but rather as a characterization of abstract features of spacetime regions, those that we may capture by considering the reduced density matrices associated with these spacetime regions. Like the ontic structural realism of Ladyman and French, this interpretation of quantum ontology has been defended for more accurately capturing the way physics represents the world than does wave function realism. Although spacetime state realists don't complain that wave function realists force us, in adopting the existence of the wave function, to address metaphysical questions underdetermined by the physics. Rather, spacetime state realists have argued that there is a more basic problem with the position. The ontology the wave function realist wants to promote for quantum mechanics, one of a field (or a field and marvelous point) evolving on a high-dimensional space with the structure of a classical configuration space, is adequate only for toy versions of quantum mechanics. And there is no suitable ontology the wave function realist could provide that would be adequate to capturing the mathematical structure of less idealized quantum theories, including quantum field theories.

Chapter 4 will address this concern, by first rebutting the arguments that in principle, wave function realism cannot be so extended, and second by sketching how wave function realism might look in the context of less idealized quantum theories. The purpose of the present chapter is not to defend wave function realism against attacks, but instead (recall) to demonstrate why the argument for wave function realism from entanglement fails to be successful, by showing that there are interpretative rivals to the framework that also adequately accommodate the existence of and distinctions between entangled states. And so we should now say a bit more about the sort of picture advocated by the spacetime state realist.

As mentioned, the spacetime state realist takes the ontology of quantum theories to be one of spacetime regions assigned highly abstract properties. Wallace and Timpson explicitly reject the position of the holist or monist who would argue that fundamentally there is only the one, the cosmos, or the whole global spacetime. For they argue, such a view does not have enough structure to support the existence of the multitude of entities we find in our empirical engagement with the world:

> Note that if . . . we were to treat the universe just as one big system, with no subsystem decomposition, then we would only have a single property bearer (the universe as a whole) instantiating a single property (represented by the universal density operator) and we would lack sufficient articulation to make clear physical meaning of what was presented. (2010, p. 710)

This is a defect they also find in wave function realism. Instead, they take as fundamental multiple subsystems the various spacetime regions, each attributed their own property represented by the appropriate density matrix.

Density matrices or operators ρ are mathematical representations that are especially useful when one is considering subsystems of larger quantum systems in entangled states. They are definable in terms of a system's wave function. For example, for pure states:

$$\rho = |\psi\rangle\langle\psi|.$$

Subsystems may then be described by a reduced density matrix that captures information about the likelihood the subsystem is in one or another pure state. This is defined in terms of the partial trace operation. For example, for our joint system described by the wave function ψ_{AB}, the reduced density matrix for atom A is defined as:

$$\rho_A = \text{tr}_B(\rho_{AB}),$$

where tr_B denotes the partial trace operation. This reduced density matrix connects up with the empirical facts in the way one would hope for an adequate interpretation of quantum ontology, yielding the right statistics for measurements made on subsystem A (mutatis mutandis for the reduced density matrix for subsystem B) (cf. Nielsen and Chuang 2016, p. 107). It should be noted, however, that as the reduced density matrices are derived from the wave functions of larger systems, there is a kind of redundancy in the fundamental ontology of the spacetime state realist, one that isn't similarly found in other approaches.

2.7 Rivals: The Multi-Field Approach

The final interpretative framework we will discuss here has been defended less frequently than the others; however, it too provides a way to capture the facts of entanglement while avoiding interpreting the wave function literally as a field on a high-dimensional space. This is the multi-field approach, in which the wave function is again represented as a kind of field, but one taking values in ordinary, low-dimensional space.

As we saw, the simple field approach also interpreted the wave function as representing a collection of fields in three-dimensional space. But this natural interpretation failed to capture the facts about entanglement. For example, the simple field approach was incapable of capturing the distinction between the quantum states:

$$\psi_{AB} = \frac{1}{\sqrt{2}} \,|L1{>}_A\,|L4{>}_B + \frac{1}{\sqrt{2}}\,|L2{>}_A\,|L3{>}_B$$

and

$$\psi_{A'B'} = \frac{1}{2}|L1{>}_{A'}|L3{>}_{B'} + \frac{1}{2}|L1{>}_{A'}|L4{>}_{B'}$$
$$+\frac{1}{2}|L2{>}_{A'}|L3{>}_{B'} + \frac{1}{2}|L2{>}_{A'}|L4{>}_{B'}$$

The simple field approach interpreted both states as describing a situation in which there were two fields, one for A and one for B, the first with half of its amplitude at L1 and half at L2, the second with half of its amplitude at L3 and half at L4.

The multi-field approach avoids conflating states like ψ_{AB} and $\psi_{A'B'}$. Peter Forrest outlined this approach in 1988:

I posit polywaves, which are disturbances to polyadic fields. The familiar monowaves (monadic waves) are assignments to each location of some member of the set of possible field-values for that location. Likewise an [N]-adic polywave is an assignment to each ordered n-tuple of locations of a member of the set of possible field-values for that [N]-tuple of locations. The integer [N] is just the "number of particles." And the possible field-values are [N]-adic relations. (1988, p. 155)

More recently, the multi-field approach has been discussed in work by Belot (2012) and Norsen, Marian, and Oriols (2015), and defended in Hubert and Romano (2018). While the simple field approach posited multiple fields corresponding to each particle, the multi-field approach, like wave function realism, postulates a single field. However, this is not an ordinary kind of field (what Forrest calls a monowave). The multi-field is an assignment of values not to individual points in space but rather to collections of points. And so, in the example we have been considering, we may consider the collections of points:

(L1, L3)
(L1, L4)
(L2, L3)
(L2, L4).

In the multi-field approach, we may interpret the quantum state ψ_{AB} as describing a multi-field that assigns a value of $1/\sqrt{2}$ to the collection of points (L1, L4) and a value of $1/\sqrt{2}$ to the collection (L2, L3). We may interpret $\psi_{A'B'}$ in terms of a multi-field that assigns a value of $1/2$ to each of the four collections of points above. Thus the multi-field interpretation is also capable of distinguishing entangled from non-entangled states. And it does so while retaining a low-dimensional field ontology.

2.8 The Contingency of Wave Function Realism

To conclude this critique of the argument from entanglement, the existence of the several ontological approaches described in the previous sections undermines this way of motivating wave function realism. The primitive ontology approach, holism, monism, the view of the wave function as a property of collections of particles, relational holism, ontic structural realism, spacetime state realism, the multi-field view—there exists a robust variety of ways to spell out the ontological lessons of quantum theories, each capable of providing descriptions of how the world is according to these theories that adequately distinguish entangled states. And so it is certainly not *essential*, as Albert put it, if one wants to be a realist about quantum theories, to take the wave function seriously as a physical field—there are realist alternatives available that take it alternatively as occupying the ontological category of law, property, pattern of relations, and so on.[11] Nor is it true, as North says, that we must formulate the dynamics on a higher-dimensional space and infer from there to a higher-dimensional ontology. Of the rival approaches considered, many interpret the wave function as something evolving on a space of lower dimensions, and those that do represent the wave function in a higher-dimensional space provide an interpretation according to which it is not necessary to view that space as physical. Finally, I must reject the flippancy of my own earlier comments that entangled states can only be distinguished, and hence completely characterized, in a higher-than-three-dimensional configuration space. It should by now be clear that there are many alternative ways to do so.

Backing up a bit, North might challenge my claim to have adequately represented her point. For the principle she appeals to in giving a version of what I am calling the argument from entanglement she also spells out in the following way:

> The rule to infer the space-time structure needed for the dynamical laws comes in two parts. First, we don't infer more space-time structure than what's needed for the dynamics. . . . We infer the least space-time structure to the world that's needed to formulate the fundamental dynamics. Any additional structure is excess,

[11] More recently, Albert has argued for wave function realism based on considerations of separability, moving beyond the point made in his earlier works. See Albert (manuscript).

superfluous structure, not in the world. . . . Second, we infer at
least as much structure as needed for the dynamics. (2013, p. 187)

Thus, North might object to the alternative approaches to interpretation I've just sketched that they either invoke too much or not
enough structure than is needed to provide an ontological support
for the dynamical laws.

However, I don't believe this is a correct diagnosis of the problem
with the rival approaches I just presented. In most of the approaches
we have just considered, there is no additional ontology being
considered that one might regard as superfluous.[12] Nor is there any
less. The wave function is simply being reinterpreted. Rather than
a field on a higher-dimensional space, the wave function is being
thought of in some other way, as a property, law or set of relations,
as a representation of a three-dimensional fundamental structured
whole, or as a multi-field on a low-dimensional space. This doesn't
mean we cannot evaluate these rival interpretations and view them
in some way or other as better or worse than the way of viewing
the wave function that the wave function realist defends. I will do
so in Chapter 3. But the problem will not be that these approaches
are not capable of accommodating the distinctions we need to draw
between empirically distinct quantum states without adding too
much superfluous structure.

The case of the primitive ontology approach is different since
more than the wave function is postulated. However, here too, it
does not seem correct to say that the defender of this approach fails
North's principle by positing too much structure to accommodate
the dynamical laws. For as we've seen, defenders of the primitive
ontology approach propose reinterpretation of the dynamical laws
themselves. Now one might try to argue that they are wrong to
think one needs to modify the laws in this way, as I will argue below.
But the problem with these approaches isn't that they fail North's

---

[12] I did note, however, that spacetime state realists appear to posit excess structure.

principle of ontological commitment and so fail to adequately accommodate the facts of entanglement encoded in the fundamental dynamical laws without superfluous structure. Given the dynamical laws they take to make up quantum mechanics, they posit exactly the right amount of structure.

Now let me be clear. To say that there are many approaches in addition to wave function realism that adequately capture and distinguish entangled states is not to say that any of these approaches are adequate to capturing *the full range of* quantum states. This would require that the ontological framework in question have the capacity to provide clear interpretations that distinguish the kind of quantum states that appear in the full range of quantum theories physicists have developed. Chapter 4 will examine what may be said about some of these frameworks' ability to provide an account of the world according to relativistic quantum theories, including quantum field theories. For now, we are just considering the more basic question of whether there are frameworks that can handle some of the states we see in nonrelativistic quantum mechanics. There is consensus that wave function realism can be used to generate a clear interpretation of these kinds of states, as can several other frameworks. The question then is what reason there is to prefer wave function realism over these other approaches to addressing the ontological question for nonrelativistic quantum mechanics. The next chapter develops an answer to this question.

3

The Virtues of Separability
and Locality

3.1 The Case for Wave Function Realism

It should now be clear that there are several ways of answering the question of what quantum theories are suggesting about the underlying nature of our world. Moreover, these interpretative frameworks all appear capable of capturing and distinguishing entangled states of the kind we find in nonrelativistic quantum mechanics. And so one cannot simply move from the claim that wave function realism is capable of capturing the facts of entanglement, to the claim that it provides an accurate description of what our world is like.

But even if the facts of entanglement underdetermine an answer to the ontological question, this does not mean a case cannot be made for preferring wave function realism over rival frameworks. In this chapter, I present such a case, citing both theoretical and pragmatic reasons for preferring wave function realism over other ontological frameworks. The core claim will be that wave function realism is unique in yielding pictures of the world with two intuitively nice metaphysical features: separability and locality. Although the pictures of the world yielded by the primitive ontology approach, holism, structuralism, spacetime state realism, and the rest may be made empirically adequate (at least for the nonrelativistic case), they fail to provide pictures of the quantum world that are fundamentally both separable and local. Since, as

The World in the Wave Function. Alyssa Ney, Oxford University Press (2021). © Oxford University Press.
DOI: 10.1093/oso/9780190097714.003.0003

I will explain, having separable and local interpretations of our best physical theories yields both theoretical and practical benefits, this is a reason to favor wave function realism as an approach to interpreting quantum theories.

As we will see, that the pictures of the world offered by wave function realism are both separable and local *in the senses that matter* is not a completely straightforward issue. To provide an argument for wave function realism from considerations of separability and locality, it will be necessary to distinguish several senses of what it could be for an ontology to be separable or local, and also to evaluate what are the virtues of an ontology's possessing these features in the relevant senses. In the end, I will argue that the pictures of the world yielded by wave function realism are fundamentally separable and local in senses that make wave function realism worth considering and developing in greater detail. This is not to say that other interpretations do not have other virtues that give us reasons to consider and develop them as well. But, I will argue, the argument from separability and locality does provide reason to prefer wave function realism to these other frameworks.

Let's begin this work of building a case for wave function realism by getting clearer on the sense in which an ontology for a physical theory may be separable. We will then turn to the other issue of locality.

3.2 Separability

Separability is a feature of physical systems that are spread out in such a way as to have constituents that individually occupy distinct regions. A system located at a region R is separable when it consists of subsystems located at nonoverlapping proper subregions of R each possessing their own individual states, and all states of the system at R are wholly determined by or grounded in the states of those subsystems. Or, to put it another way, as Howard writes:

[Separability] is a fundamental ontological principle governing the individuation of physical systems and their associated states, a principle implicit in many classical physical theories. It asserts that the contents of any two regions of space-time separated by a nonvanishing spatiotemporal interval constitute separable physical systems, in the sense that (1) each possesses its own, distinct physical state, and (2) the joint state of the two systems is wholly determined by these separate states. (1989, pp. 225–226)

The key differences between my formulation and Howard's are two. First, his includes the constraint that the relevant subsystems are those separated by some "nonvanishing spatiotemporal interval." It is not clear why this should be required so long as the subsystems fail to overlap spatially. Second, I don't speak of spacetime regions in order that we may be neutral about the background framework in which these systems and subsystems inhere. We may, as is common, consider objects and their parts existing in ordinary three-dimensional space or spacetime. But we may also consider what is to be said about the ontologies proposed by the wave function realist, ontologies spread out against different, higher-dimensional backgrounds.

One drawback to such standard ways of construing separability is that, as they stand, they seem to require that a separable metaphysics be a Humean metaphysics.[1] Since they impose the requirement for separability that *all* facts be determined by facts about what occurs at individual subregions, this appears to require that for a metaphysics to be separable, all facts about dispositions, counterfactuals, causation, and laws must be determined by what occurs at individual regions.[2] One might avoid this implication by modifying the definition of 'separable' so that it only states that

[1] See Section 3.9 for a clear statement of the central doctrines of Humean supervenience.
[2] As we will see, Loewer (1996) defends wave function realism explicitly for its ability to provide an interpretation of quantum theories compatible with Humean supervenience.

categorical, that is, nondispositional or non-nomic facts about joint regions are determined by the facts about individuals at subregions. We then have the following.

> A metaphysics is *separable* if and only if (i) it includes an ontology of objects or other entities instantiated at distinct regions, each possessing their own, distinct states, and (ii) when any such objects or entities are instantiated at distinct regions R1 and R2, all categorical facts about the composite region R1∪R2 are determined by the facts about objects and properties instantiated at R1 and R2 individually.

As we will shortly see, entanglement can appear to force on us the violation of even this weaker account of separability.

The notion of separability (and nonseparability) is easily illustrated using nonscientific examples. For example, that a pair of tennis balls is multi-colored is a separable state of the pair, for that the balls together are multi-colored may be determined by the colors of the individual balls, say, one being orange and the other yellow, or by the fact that the individual balls have spatial parts, some which are one color and others another color. On the other hand, a couple's being married is a nonseparable feature of a pair of individuals, technically, since it is not determined by the states of the individuals taken separately. A couple's being married is not a particularly interesting nonseparable feature, since although it is true that it is not determined by intrinsic features of some individuals taken separately, it is determined by the state of those individuals plus the features of some other things in the couple's environment. And so we may say that there is a larger system, one that includes the couple's environment, for which separability obtains. The kind of nonseparability suggested by quantum entanglement is rather more interesting. In these cases, there appear to be features of composite systems that resist determination by features of individual subsystems, including the subsystems making up those systems' environment. Indeed, in cases of quantum entanglement, there appear to be (categorical) features of composite systems that fail to be determined by anything more fundamental.

Systems in quantum mechanically entangled states, like Bohm's pair of atoms, are thus often thought to exhibit fundamental non-separability. Recall the spin state of Bohm's atoms, the singlet state:

$$\psi_{singlet} = \frac{1}{\sqrt{2}}\,|x-up\rangle_A\,|x-down\rangle_B - \frac{1}{\sqrt{2}}\,|x-down\rangle_A\,|x-up\rangle_B$$

The singlet state is thought to be an example of a nonseparable state because when a pair of atoms is in such a spin state, this fact is not determined by any facts about the atoms individually, including facts about their individual x-spins. For atoms in the singlet state, it is a fact about the pair that if their individual x-spins were measured, it is absolutely certain that they would be found to have opposite spins. But in such a state, there is no definite fact about the atoms' individual x-spins that could serve to determine this fact about their relative spins. For each atom, we know only that if we conduct an x-spin measurement, there is a 50% chance of finding an x-up result and a 50% chance of finding an x-down result. We know nothing more definite. Schrödinger emphasized this seeming consequence of quantum mechanics explicitly in 1935:

> Maximal knowledge of a total system does not necessarily include total knowledge of all of its parts, not even when these are fully separated from each other and at the moment are not influencing each other at all. (1935, p. 160)

Systems can be and often are in entangled states of many variables, such as spin, position, momentum, and energy. And nonseparability has seemed to be a pervasive feature of quantum mechanics, due to such entanglement. It is a feature we can now see is explicitly taken to be a metaphysical consequence of quantum theories according to most of the ontological frameworks we considered in Chapter 2.

This is most obvious in the relational approaches of Teller and the ontic structural realists. Relational holism just is the position that collections of objects have physical relations which fail to obtain in virtue of the non-relational physical properties of their parts. And ontic structural realism is the view that there are no fundamental intrinsic features, features that could ground the relations postulated by fundamental physics. Because facts about the density matrices of total quantum systems, including spacetime as a whole, are not determined by the density matrices applying to individual subsystems, spacetime state realists are also committed to fundamental nonseparability, and they are quite explicit about this (Wallace and Timpson 2010, Section 6). The multi-field approach is also nonseparable in taking there to be facts about field assignments to collections of spacetime points that are not determined by facts about the features assigned to any individual points. And, in postulating properties of collections of particles that are not determined by any properties of their constituent individuals, wave function-as-property approaches are similarly nonseparable, as is priority monism, which postulates relations between spatially separated particles that are determined not by these particles' individual features but only by features of the one, the cosmos. Finally, separability fails as well on the existence holist view Howard took to be a consequence of quantum entanglement. In this case, separability fails, however, not by violating the second clause of the definition, but the first. Recall: a system located at a region R is separable when (i) it contains subsystems located at nonoverlapping proper subregions of R each possessing their own, distinct states, and (ii) all categorical features of the system at R are wholly determined or grounded by the local features of those subsystems. The existence monist rejects the very existence of subsystems at the distinct subregions and so fails to have a separable ontology in that respect.

Momentarily we will discuss how the wave function realist reveals the apparent nonseparability manifested by quantum mechanically entangled systems to be a consequence of a more fundamental metaphysics that is separable in higher dimensions. But it is worth mentioning beforehand that there is another way of ensuring separability, while accepting the reality of quantum entanglement. That is by adopting the primitive ontology approach to quantum theories. As we've seen, the interpretations that fall out of the primitive ontology approach when it is applied to the various solutions to the measurement problem always include a dual ontology of (a) local beables spread out in space or spacetime, each possessing certain determinate features, for example, particles with determinate values of position, matter-density fields taking on determinate values at points, or GRW flashes possessing determinate spacetime locations, and (b) a wave function, sometimes understood as a physical wave in a high-dimensional space, a so-called guiding wave, but more often as something with a nomological status, determining like a law how the primitive ontology behaves over time (Goldstein and Zanghì 2013, Allori forthcoming).

By adopting the primitive ontology approach and taking the wave function to be something with nomological status, one might argue that there are no facts about joint states of local beables at distinct spatial or spacetime regions that fail to be determined by the states of these beables taken individually. In cases of quantum entanglement, there will generally be facts about something else, the wave function, that is not determined by the states of the local beables (taken individually or together). However, one could say that the primitive matter ontologies are perfectly separable. Since the wave function has a nonprimitive status, facts about it fail to provide counterexamples that would undermine claims of separability. And so, unlike the majority of rivals to wave function realism, the primitive ontology approach does appear capable of yielding separable metaphysics. Perhaps because of the way this separability is achieved, by considering the wave function to have a nonprimitive status, these ontologies also tend to be nonlocal.[3]

3.3 Separability and Wave Function Realism

The key innovation of wave function realism is to postulate the existence of a field that is spread out not in the three-dimensional space of our ordinary experience, but instead in the space of the universal wave function. This will be a higher-dimensional space, which, for nonrelativistic quantum mechanics, has the structure of a classical configuration space. So, unlike the primitive ontology approach, which takes as fundamental both ordinary low-dimensional matter and a wave function, in this picture, fundamentally, there is only what inhabits the high-dimensional space of the wave function. The matter itself is viewed by the wave function realist as constituted out of the wave function (and perhaps a marvelous point). The resulting wave function metaphysics is capable of capturing entanglement relations while remaining completely separable. It is separable because all states of the wave function, including the entangled states we have been considering, are completely determined by localized assignments of amplitude and phase to each point in the higher-dimensional space of the wave function.

To illustrate, let's return to Bohm's example involving two entangled atoms sent through two Stern-Gerlach devices so that their paths are bent by the magnets according to their x-spins. The wave function realist argues that one should take quantum

---

[3] The view spelled out by Travis Norsen in his 2010 "The Theory of (Exclusively) Local Beables" also explicitly has this character. Norsen works in the context of Bohmian mechanics to develop a metaphysics that is completely separable, containing only particles with definite positions and three-dimensional fields. But to have an empirically adequate, separable ontology for quantum mechanics in low dimensions, the picture is forced to be (causally) nonlocal.

mechanics to be suggesting that these atoms are ultimately constituted out of something inhabiting the wave function's higher-dimensional space. What exactly this is depends on the best way to solve the measurement problem. If this is Bohmian mechanics, then quantum mechanics is suggesting the atoms are constituted out of the wave function and a marvelous point. If the best way to solve the measurement problem is Everettian quantum mechanics or a collapse theory, then quantum mechanics is suggesting the atoms are constituted out of the wave function alone.

As I argued in Chapter 1, the motivations for pursuing a hidden variables theory like Bohmian mechanics makes wave function realism rather a bad fit. For this reason (in addition to a desire to keep the discussion that follows manageable), I will tend to focus on how the wave function realist understands the underlying ontology of Bohm's scenario in the context of Everettian and collapse versions of quantum mechanics.

According to both approaches, at the time when the atoms are initially deflected by their magnets on their paths to their respective measurement screens, it is very likely that they remain in an entangled position state. For the Everettian who only recognizes Schrödinger evolution, this is certain. Spontaneous collapse theories will predict that at some point, likely upon interaction with the particles in one of the detection screens, the system will undergo a hit. But before this interaction takes place, the system will remain in an entangled state of position with high amplitude around the two points in wave function space we earlier suggestively labeled L14 and L23.[4] The entanglement is thus completely captured in the wave function realist's ontology by localized facts about individual locations: there is substantially higher wave function amplitude around L14 and L23 than around L13 and L24. To capture the facts of entanglement, there is no need for the wave

---

[4] See Albert and Vaidman (1989) for a treatment of the Stern-Gerlach experiment using the GRW framework.

function realist to postulate any irreducible relations between spatially distant objects.

Thus, the wave function metaphysics is fundamentally separable, in the sense that there are no categorical facts about spatially separated objects that are not determined by local facts about the wave function and its parts. Fundamentally, all that exists according to quantum theories on a wave function realist interpretation is the wave function evolving in its own high-dimensional space. There are no relational facts that fail to be determined by local facts about the wave function.

This, however, is not all the wave function realist has to say about quantum nonseparability. For most wave function realists think that, in addition to whatever fundamental ontology is implied by quantum theories, we can also speak about a less fundamental, derivative reality. We can also make true claims about a world of objects spread out in three-dimensional space or spacetime. And this raises an additional question about separability: does separability obtain as well in the low-dimensional derivative pictures of the world also accepted by the wave function realist?

The simplest thing for the wave function realist to say in response to this question is, "No, the derivative three-dimensional reality is nonseparable." After all, it is true that relational facts about derivative, low-dimensional systems (e.g., the fact that atoms A and B will always be found to have opposite spin) do not obtain, according to the wave function realist, in virtue of localized facts about their individual *low-dimensional* subsystems.

Now although the wave function realist concedes that such derivative nonseparability obtains, this is not to say that any low-dimensional relational facts are fundamental facts, or fail to have further explanations. The wave function realist takes such relational facts to be explained in terms of the features of a more fundamental ontology that is perfectly separable. Because all such relational facts ultimately have explanations, there is no cause for discomfort with this form of nonseparability. Nonseparability of this derivative kind doesn't indicate strangeness in the more fundamental reality that produces what we observe.

As an analogy, consider the kind of strangeness one might find in certain special contexts, such as when one is playing an immersive video game, engaging with a virtual reality. In this immersive environment, strange things may happen, objects may behave in ways we might not otherwise expect. However, all of this strange behavior may be ultimately explained by a more fundamental reality that does not have these features. The derivative appearances that manifest in the environment in which we are immersed need not be a simple reflection of the features of the more fundamental reality.[5]

3.4 A Challenge

In recent work, Myrvold (2015) and Lewis (2016) have challenged the claim that wave function realism provides a picture of the world in which all facts are determined by an assignment of local values to points or subregions of the wave function's space. In other words, they challenge the separability of the wave function realist's proposed metaphysics. If they are right, then this is a serious concern, as I argue that the best case for wave function realism involves the framework's ability to yield pictures of the world that are fundamentally separable (and local). In this section, I present and respond to their concerns, which are found in Myrvold's 2015 paper "What Is a Wavefunction?"

According to Myrvold, it fails to be the case that facts about the wave function at arbitrary subregions of its space are local facts about these regions. When we say the amplitude of the wave function is high at L14 or low at L13, these are nonlocal facts. And so

---

[5] To say that the derivative reality is like a virtual reality or immersive video game is not necessarily to say that the objects in it are not real or that they do not really possess the properties they appear to. See David Chalmers (2017) for a defense of virtual realism.

there is no basis of localized facts upon which one might rest a claim of separability.

To see the nature of this concern, Myrvold asks us to consider two states. S1 is a state in which a single-particle wave function takes on a nonzero value at some point x. S2 is a state just like the first near x, but differs in that there is with certainty a particle confined to a spacetime region R, far from x. It follows by construction that in S2, the probability that an array of detectors spread through space will report a particle detection at x and nowhere else is zero. This means that the single-particle wave function for S2 must take the value zero at x. A nonzero value of a single-particle wave function at x is incompatible with there being a particle definitely located in region R, no matter how far away R is. Myrvold's conclusion is that the assignment of the nonzero value to some point near x in S1 is therefore not a local fact about that point. And from this he infers that wave functions are not separable, despite the claims of the wave function realist. He also argues that this fact makes them not like traditional fields, since traditional fields are defined by a local assignment of features to points. And so, Myrvold's argument, if successful, would go further than merely undermining the wave function realist's claim to provide fundamentally separable pictures of the quantum world. It would also undermine a central aspect of their view about wave functions.

Myrvold's argument may be summarized in the following way:

1. That a single-particle wave function takes on a nonzero value at a point x is incompatible with there being a particle definitely located in a region R that does not overlap x, no matter how far R is from x.

2. So, facts about the assignment of values of wave functions to regions of wave function space depend on circumstances in distant regions.

3. So, wave functions are not defined by an assignment of local quantities to regions of the space they inhabit.

4. So, the wave function realist's picture of the world is not fundamentally separable.

5. And, wave functions are not fields in the usual sense.

I wish to argue that depending on what one means by "assignment of local quantities," the wave function realist either has good reason to reject the inference from (2) to (3) or she has good reason to reject the two inferences from (3) to (4) and (3) to (5).

A natural understanding of what is meant by an assignment of a local quantity is an assignment of a feature that is not determined by what takes place at distant locations. Assuming this understanding, it is easy to see that (3) does not follow from Myrvold's earlier claims. What we find in S1 and S2 are two distinct wave functions, where the state of each wave function is determined by the assignment of amplitudes (and phase, though this is ignored in the example) individually to each point in its space. The amplitude of the single-particle wave function at x in S1 is not equal to the amplitude of the single-particle wave function at x in S2 and this is what determines the fact that these are distinct wave functions. Now Myrvold notes that we can infer that the single-particle wave function's amplitude at x for S2 differs from that of S1 simply by knowing that in S2 there is a particle definitely located at a distant region R. This is due to a modal dependence between the fact about the particle and the fact about the wave function. But, that there is a particle definitely located at R is not what metaphysically determines the amplitude of the single-particle wave function at x for S2 according to the wave function realist. The facts about the amplitude of the wave function at different points in its space are brute. According to the wave function realist, facts about the locations of particles are ontologically derived from facts about the wave function (or wave functions). The fact about dependence stated in (2) obtains due to this fact of ontological priority. So, (3) simply does not follow from (2). This part of Myrvold's argument simply begs the question against the wave function realist. It assumes that facts about the locations of particles ontologically determine facts about wave functions rather than vice versa.

Now Myrvold would likely reject this understanding of what makes for an assignment of local quantities. Myrvold himself makes the following terminological remarks helpful to interpreting his argument:

> We will say that a physical quantity is *intrinsic* to a spacetime region if the fact the quantity has the value that it has carries no implications about states of affairs outside the region. A *local beable*, as we understand it, is one that can be regarded as an intrinsic property of a bounded spacetime region, and will be said to be local to that region. (2015, p. 3251)

So, let us evaluate the argument with this understanding in mind. In S2, since we are told that a particle is definitely located at R, we must infer that the single-particle wave function takes on value one at location R. Assuming the wave functions Myrvold describes are normalized, this would mean that S2's single-particle wave function must take on a value of zero at every other location. From this, we can infer that its amplitude at x is zero. Now one might read Myrvold as saying that it is this fact that demonstrates that the assignment of amplitudes to the wave function is not an assignment of local quantities. The facts about the assignment of wave function values at R carry implications for the assignment of wave function values at x.

Under this second understanding of "assignment of local quantities" then, (3) follows. But why should this make the wave function metaphysics nonseparable or the wave function unfield-like? Start with the latter question. Consider paradigmatic examples of entities typically represented as fields with amplitude and phase: bodies of water, waves on a string. Since actual water waves or waves on a string are always made of a finite amount of material, there necessarily will be implications between the amplitude the wave takes on at one location and those it takes on at others. So, in Myrvold's sense, one would say that waves of these kinds are not defined by an assignment of local features because there are dependence relations obtaining between the values the wave takes on at distinct locations. So noting that the wave function too has this feature does not make it un-field-like unless all of these things are un-field-like as well. And so even if the move to premise (3) is valid, the further move to (5) is not.

We may say the same for the inference to conclusion (4). Whether the wave function metaphysics is separable is a matter of whether facts about features at composite spatial regions are determined by facts about features at smaller regions. Recall the earlier account of separability, and that of Howard who interprets it as

> a fundamental ontological principle governing the individuation of physical systems and their associated states. . . . It asserts that the contents of any two regions of space-time separated by a nonvanishing spatiotemporal interval constitute separable physical systems, in the sense that (1) each possesses its own, distinct physical state, and (2) the joint state of the two systems is wholly determined by these separate states. (1985, pp. 225–226)

Again, separability is a matter of determination, not dependence. Separability may obtain even if there is a dependence in values taken on by the wave function at distant regions. Such dependence has no bearing on whether what happens at a composite region is determined by what happens at the regions out of which it is composed. And so, when we accept (3) using Myrvold's interpretation, (4) does not follow.

Oliver Pooley (personal communication) has noted that if one was motivated to adopt a separable, wave function metaphysics out of Humean concerns—for example, as Loewer has advocated in his paper "Humean Supervenience"—then one might be concerned that my responses to Myrvold's argument don't save the view after all. We will discuss this argument of Loewer's in more detail a bit later in this chapter. For now, let's concede that a (the?) core Humean doctrine is that there are no necessary connections between distinct existents. And yet, I am allowing here that there may be strong modal connections between the value the wave function takes on at one point in its space and another. I say "strong modal connections" here because I am unsure whether the wave function realist should take it to be a necessary fact that wave functions are normalized. I think that in worlds similar to ours, in the sense that they are worlds that support the existence of familiar observable objects as derivative entities, wave functions will be normalized and so there will be correlations between values in different parts of space. But this doesn't entail that normalization must hold in all possible worlds. At least in Everettian quantum mechanics, there is no contradiction in the idea of a world with a non-normalized wave function. Though in GRW, with the ordinary collapse law, if the wave function is not normalized, the collapse probabilities would not sum to one. This seems close to a contradiction.[6] Nonetheless, there do appear to be strong (even if not necessary) modal correlations between the values the wave function takes on in different parts of its space, and these appear to be in violation of the spirit of Hume's doctrine.

To reply to this concern, I would again point to the distinction between determination and dependence. I believe that the Humean's concern ought to be with determination between distinct existents, not simply dependence. Determination between distinct existents in distinct parts of space is not part of the wave function realist's picture. Any correlations and hence dependence stem from brute and local facts about the state of the wave function. It is not the case that what is going on in one part of the wave function's space metaphysically determines what is happening in another. Humeans allow there to be correlations between distinct existents in general.

[6] Thanks to Chip Sebens here.

For example, if all electrons have negative charge, this is such a correlation. It doesn't come from the charge of one electron determining the charge of another. It just comes from a brute fact about the state of all of the electrons. The correlations under discussion here similarly come from a brute fact about the state of the wave function in our world.

Thus I take there to be no compelling objection to the wave function realist's claim that the picture of the world she considers quantum theories to offer is fundamentally separable. With that established, we may now move on to discuss the second nice metaphysical feature the wave function realist claims her interpretations of quantum theories have: locality.

3.5 Concepts of Locality

The issue of locality has frequently been conflated with that of separability in the scientific and philosophical literature. This is not really so surprising given the multiplicity of meanings our language assigns to "local." But unlike separability, which concerns the noncausal *metaphysical* determination of the features of a total system by the features of the subsystems located at that system's subregions, locality, in the sense to be discussed here, is a causal notion, tracking facts about the *causal* determination of events. Some philosophers find it helpful to think in terms of a contrast between vertical and horizontal determination relations. Separability concerns the status of vertical determination relations between more and less fundamental entities; locality the status of horizontal determination relations between causes and effects.

The concepts of locality most frequently invoked when quantum entanglement is under discussion are those highlighted by John Bell. Howard Wiseman (2014) has argued that Bell really had two different accounts of what locality may come to in physics. The first is the notion of locality invoked in his 1964 paper "On the Einstein-Podolsky-Rosen paradox" which we will discuss in more detail momentarily:

> the requirement of locality, or more precisely that the result of
> a measurement on one system be unaffected by operations on a
> distant system with which it has interacted in the past.

This is often taken to be equivalent to what Abner Shimony (1990) called "parameter independence." Applied to the case of Bohm's entangled atoms, it is the principle that the probabilities for the results of a measurement on atom B are independent of what we choose to do to the spatially separated atom A, including what measurements we choose to perform on it (and mutatis mutandis for the results of measurements on A). In a relativistic setting, and these principles are usually put forward in a relativistic context, the claim is that the probabilities for the results of a measurement on B are independent of what measurement we choose to perform any spacelike distance away. Two events are spacelike separated from one another when it would take a signal traveling faster than the speed of light to reach one from the other.

In his 1976 paper, "The Theory of Local Beables," following Wiseman, we may see Bell as invoking a distinct principle that he calls "local causality":

> Let A be localized in a space-time region 1. Let B be a second
> beable in a second region 2 separated from 1 in a spacelike
> way.... Now my intuitive notion of local causality is that events in
> 2 should not be "causes" of events in 1, and vice versa.

By "beable," as we've seen, Bell simply means entity, something that is real. Local causality is a stronger principle than the earlier "locality" principle from 1964. It states not only that the probabilities for the results of a measurement on one system are independent of how we may manipulate another system at a spacelike separation from it, but also that these probabilities are independent of the actual measurement results we find when we measure that other system. Wiseman argues that it is local causality that Bell took to be the primary locality principle of interest from at least 1976 on. And, he argued, it is a principle that must be violated if quantum theory is to be correct.[7]

3.6 Quantum Nonlocality

Again, my goal in this chapter is to show that wave function realism has an advantage over other frameworks in offering pictures of the world that are fundamentally both separable and local. But to be clear how and why the locality of a wave function ontology gives wave function realism an advantage over other ontological frameworks, we must pause to see why the quantum world is often thought to be nonlocal in the first place.

That quantum mechanics may potentially, under some interpretations, lead to nonlocality was first shown in the famous 1935 paper by Albert Einstein, Boris Podolsky, and Nathan Rosen (EPR), "Can Quantum-Mechanical Description of Physical Reality Be Considered Complete?" In that paper, Einstein, Podolsky, and Rosen don't really take seriously the possibility that our world might be nonlocal. They rather assume that spacelike separated events do not influence each other and use this assumption, among others, to argue that the quantum (i.e. wave function) formalism should be supplemented with a description of additional variables of the kind we find in, for example, De Broglie's pilot wave model. Their conclusion is stated thus:

---

[7] There is some debate in the literature about whether these are really distinct principles. To my mind, Wiseman makes a compelling case. But see also Norsen (2015).

Previously we proved that either (1) the quantum-mechanical description of reality given by the wave function is not complete or (2) when the operators corresponding to two physical quantities do not commute [i.e., are, in Bohr's terminology, complementary] the two quantities cannot have simultaneous reality. Starting then with the assumption that the wave function does give a complete description of the physical reality, we arrived at the conclusion that two physical quantities, with noncommuting operators, can have simultaneous reality [i.e., they have arrived at a paradox]. Thus the negation of (1) leads to the negation of the only other alternative (2). We are thus forced to conclude that the quantum-mechanical description of physical reality given by wave functions is not complete. (1935, p. 780)

By saying that a theory is complete, Einstein, Podolsky, and Rosen mean that the theory does not leave out any element of physical reality (1935, p. 777). They do not define what they mean by "having reality," but they do give the following criterion for how we may tell when it is the case that some quantity has reality in the sense they have in mind:

If, without in any way disturbing a system, we can predict with certainty (i.e. with probability equal to unity) the value of a physical quantity, then there exists an element of physical reality corresponding to this physical quantity. (1935, p. 777)

With these comments in hand, we may then summarize their argument.

For ease of presentation, I will use the example of David Bohm's two atoms with which we have been working. Bohm in fact uses this example to illustrate the EPR thought experiment and so it is standardly referred to as the EPRB set-up. Suppose the quantum mechanical description of reality is complete and our atoms begin in a spin state that is entangled with respect to their x-spin values, for example, the singlet state:

$$\psi_{singlet} = \frac{1}{\sqrt{2}} \left| x - up \right\rangle_A \left| x - down \right\rangle_B - \frac{1}{\sqrt{2}} \left| x - down \right\rangle_A \left| x - up \right\rangle_B$$

We now consider sending these atoms away from each other, and thereafter measuring only the spin of the first atom A. This, in orthodox quantum mechanics and collapse approaches more generally will serve to collapse the wave function of this system onto a determinate value of x-spin. Suppose atom A is found to have spin up in the x-direction. Then, it would seem that without in any way disturbing the other atom B, we can predict with certainty that a measurement on it will find it x-spin down. And so B's x-spin, it follows from the reality criterion, is an element of physical reality that exists even before B is measured. But, Einstein, Podolsky, and Rosen note, we have in no way disturbed atom B and so, according to quantum mechanics (with a tacit locality assumption in place, as well as the assumption that the quantum mechanical description is complete), B must still be in the spin state it started out in, a spin state that is indeterminate with respect to x-spin. So is the x-spin of atom B indeterminate as quantum mechanics would seem to require or determinately down as the inference from the measurement on A and the criterion of reality would suggest? They note:

> Since at the time of measurement [on A] the two systems no longer interact, no real change can take place in the second system [B] in consequence of anything that may be done to the first system. This is, of course, merely a statement of what is meant by the absence of an interaction between the two systems. Thus, *it is possible to assign two different wave functions . . . to the same reality.*(1935, p. 779)

Quantum mechanics, at least of the orthodox collapse variety, seems to lead to an internal tension about the states of systems when it is assumed to give a complete representation of physical reality. To resolve this paradox, EPR ask the reader to take seriously the possibility that the wave function description is not complete and should be supplemented with additional variables taking on determinate values, thus accommodating the reality of atom B's x-spin given that it will certainly be found upon measurement to have x-spin down.

Although in 1935 it may have been reasonable to think that supplementing the existing quantum mechanics with the postulation of additional, so-called hidden variables would help avoid nonlocality and paradox, subsequent theoretical and experimental results proved otherwise. Bell's theorem of 1964 and the Kochen-Specker theorem of 1967 both showed that no simultaneous assignment of determinate values to variables before measurement could yield a set of predictions for the results of measurements that were compatible with the existing quantum mechanics.[8] One could have at this time taken these results to show that quantum mechanics wasn't merely incomplete, it was wrong, but these quantum predictions were then confirmed in a series of experimental settings, most famously by Alain Aspect and his collaborators (1981). And this suggested to many that EPR were wrong about there being a paradox. A measurement on atom A could change the probabilities of measurement for a distant atom B with which it is entangled. Yet this correlation cannot be explained by any assignment of preexisting spin values for the atoms. What needs to be rejected instead of the completeness of quantum mechanics is the assumption that the quantum world, our world, is local.

[8] To be clear, it is not that Bell's theorem and that of Kochen and Specker showed there couldn't be hidden variables, so that a theory like Bohmian mechanics was inconsistent. They only showed that the postulation of hidden variables would not suffice to avoid nonlocality in the way that EPR speculated. Advocates of Bohmian mechanics generally acknowledge that quantum mechanics shows the world to be nonlocal.

To give a better idea of these results, I will present Bell's proof in outline.[9] A key feature of Bell's set-up is that the particular spin variable measured on the A and B wings of the experiment is allowed to vary. Our Stern-Gerlach magnets may be set at various orientations so that they are measuring the spin of atoms A and B along various, different axes. For atoms in an initial entangled spin state, such as the singlet state, quantum mechanics predicts (and these have been confirmed in subsequent realizations of the experiment) stable statistical correlations in the values that will be found for the spins of A and B along these various axes.

Bell argued in his 1964 paper that if it were the case that the results of measurements on a pair of entangled atoms like A and B depended only on the settings of the magnets local to A and B, respectively (he calls these settings **a** and **b**), and some more complete specification of the initial spins in the form of additional variables beyond the system's wave function, then these results over time would not conform to the predictions made by quantum mechanics. In Bell's words:

> The paradox of Einstein, Podolsky, and Rosen was advanced as an argument that quantum mechanics could not be a complete theory but should be supplemented by adding additional variables. These additional variables were to restore to the theory causality and locality. In this note that idea will be formulated mathematically and shown to be incompatible with the statistical predictions of quantum mechanics. It is the requirement of locality, or more precisely that the result of a measurement on one system be unaffected by operations on a distant system with which it has interacted in the past, that creates the essential difficulty. (1964/1987, p. 14)

---

[9] One can find excellent treatments of the details of the proof with intuitive examples not only in Bell's original paper but also in David Mermin's "Quantum Mysteries for Anyone" as well as Tim Maudlin's *Quantum Non-Locality and Relativity*.

Bell begins his argument considering the case of atoms A and B in the singlet state. The atoms are sent away from each other and the spin of each is measured using separate Stern-Gerlach devices whose magnets may be set to various orientations. He then states an assumption that the outcome of the spin measurement on atom A depends only on the setting of its magnet **a** and on the value of some additional variables λ added to the wave function. Similarly, the outcome of the spin measurement on atom B will be assumed to depend only on the setting of its magnet **b** and on values of the additional variables λ. This is to assume that the outcome of a spin measurement on one atom doesn't depend on what happens at a distant location to the other atom; in other words, this is to assume locality.

Bell then shows that with this assumption and what quantum mechanics entails about systems in states like the singlet state— for instance, that in cases in which the two magnets are set to measure identical spin components, the resulting measurement outcomes for A and B will show them to have opposite spin—he derives a relationship that, given his assumptions, must obtain between the values of the measurement outcomes. This is the *Bell inequality*, which he notes can be generalized from the case he considers. The inequality establishes a relationship between the pattern of measurement results that would have to obtain were the results of the measurement on A determined only by the local setting on its magnet **a** and the hidden parameters λ and not on what happens to B, and were the results of the measurement on B determined only by the local setting on its magnet **b** and the hidden parameters λ and not what happens to A. The important point is that the obtaining of this inequality is incompatible with the predictions of quantum mechanics, nor does it approximate these predictions. And as the quantum predictions have been confirmed, we can now know that the inequality does not obtain in nature. So, one of the assumptions leading to the inequality has to go. Bell concludes:

> In a theory in which parameters are added to quantum me-
> chanics to determine the results of individual measurements,
> without changing the statistical predictions, there must be a
> mechanism whereby the setting of one measuring device can
> influence the reading of another instrument, however remote.
> (1964/1987, p. 20)

He argues that the fault in the original assumptions leading to the
inequality was the claim that the quantum world is local. Whether
there are additional parameters added to quantum mechanics or
not, the result of a measurement on one half of an entangled pair
isn't only determined by these parameters and local measurement
settings. The conclusion he reaches is that it must also depend
on something happening a distance away. One might expect this
something to be the measurement taken on its entangled partner.

Many have followed Bell to his conclusion, that quantum mechanics requires us to see the world as fundamentally nonlocal.
However, not all have. In Section 3.7, I will describe the sense in
which the wave function realist may offer a picture of the world that
is compatible with the statistical predictions of quantum mechanics
and yet at the same time is fundamentally local. In the section that
follows (Section 3.8), I will consider alternative proposals to avoid
quantum nonlocality by appeal to fundamental nonseparability.

3.7 Locality and Wave Function Realism

Arguably, the senses of "locality" used by Bell and distinguished
by Wiseman cannot satisfactorily explicate the sense in which the
metaphysics of the wave function realist is (fundamentally) local.
For the principles invoked by Bell both concern the existence of
causal relations in spacetime, by invoking facts about certain events'
exhibiting spacelike separation. Yet there is no spacetime interval
defined on the space the wave function is said to inhabit, nor is the

space of the wave function the space in which light propagates, and so there is no sensible notion of spacelike separation in the wave function realist's more fundamental metaphysics to let us settle the issue of whether these senses of locality or local causality obtain.

To explain the way in which wave function realism provides one with local metaphysics, we must move to a concept of locality that makes sense in the context of the high-dimensional space of the wave function realist. Sometimes, when Bell discusses his principle of local causality, he states it in broader terms than we previously saw, terms that may be more useful in this context:

> What is held sacred is the principle of "local causality"—or "no action at a distance." (Bell 1981, p. 46)

This principle is generalizable so as to be of use by the wave function realist. And so we may say a metaphysics is local if and only if it contains no instantaneous action across spatial distances, where 'spatial' refers to whatever is the spatial background of the metaphysics. It may be our familiar three-dimensional space or spacetime. But it may also be the high-dimensional background of the wave function realist.

In describing the wave function realist's metaphysics as fundamentally local, we must be careful to distinguish the situation for the interpretation of different quantum theories, including those with and without wave function collapse. In the case of Everettian quantum mechanics without collapse, the wave function simply evolves unitarily in accordance with the Schrödinger equation or its relativistic variant. The wave function spreads out and may interfere with itself as waves do. But at no point does an action at one point in the space of the wave function influence an action somewhere distant in that space. So there is straightforwardly no nonlocal action.

For collapse theories like GRW, the wave function may evolve unitarily, but from time to time there is a spontaneous collapse.

This involves the entire wave function undergoing a hit, which may be represented mathematically by the multiplication of the quantum state by a Gaussian function localized within a particular region of the wave function's space, a three-dimensional subspace of the wave function's space corresponding to the possible locations of a single particle in the low-dimensional space.

In this case as well, it is not correct to say that what happens in one region of the wave function's space acts immediately to influence what happens in another. Rather, in these models, collapses are not caused by anything about the state of the wave function at the previous time, but occur spontaneously. One could say that there are facts about the wave function at the time prior to collapse that determine how likely it is that the hit is localized at one point rather than another. The probability of the collapse being localized at one point rather than another is given by the Born rule probabilities which are associated with the amplitude squared of the wave function at the different points in its space. But there is still no reason to say that the amplitude of the wave function at one distant region R causes a collapse to be localized at another region R' of the wave function's space instantaneously. Even if the wave function later becomes more peaked around R', the collapse isn't something that takes place at R', but is rather something that take place across the entire space. So there isn't really a localized effect that is influenced by some distant cause. The evolution of the wave function through collapse may be jerky and discontinuous, but it does not involve any nonlocal action.

Let's walk through this a little more carefully since it has been confusing for some to separate the nonlocality we find in the low-dimensional picture from the locality the wave function realist finds in the high-dimensional one. Suppose at some initial time t_0, our wave function is spread out in such a way as to be peaked at two locations in its space, the first corresponding to a measurement on atom A coming out as x-spin up, the second peak corresponding to a measurement on atom A coming out as x-spin down (Figure 3.1).

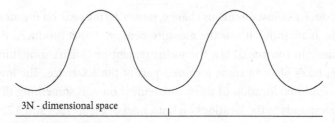

3N - dimensional space

Figure 3.1. Wave Function at t_0

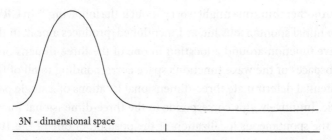

3N - dimensional space

Figure 3.2. Wave Function at t

Now suppose that at time t, A is found to have x-spin up. The wave function has thus collapsed onto a state in which the second peak has been suppressed (Figure 3.2).

Let's concede for the moment that in the three-dimensional derivative metaphysics, there is nonlocal action. The measurement on the A-wing of the experiment has caused an instantaneous change in the spin state of atom B, some distance away. However, in the 3N-dimensional, more fundamental metaphysics of the wave function realist, has some event at one location of the space caused a change in something a distance away? Has the raising of the first peak instantaneously caused the suppression of the second peak? The answer is no. The hit the wave function underwent at t caused an immediate change in the entire shape of the wave function such that it became more peaked in one region than the other. But it is certainly not the case that the peak on the one side of this diagram produced an instantaneous change, removing the peak on the other side. If anything, it was the measurement on A that produced the change in the peaks, but the measurement on A isn't something that takes place in some localized part of the 3N-space. The low-dimensional location of the measurement on A is something that corresponds to the locations on both sides of Figures 3.1 and 3.2. It occurs just as much at the location of the second peak as it does at the location of the first. If anything, the measurement occurs across the high-dimensional space.

Another thing one might worry about is the following.[10] In GRW, the initial spontaneous hit, as I mentioned, produces a peak in the wave function around a location in one of the three-dimensional subspaces of the wave function's space corresponding to all of the potential determinate three-dimensional locations of a single par-ticle. Intuitively, this corresponds in the three-dimensional image to the spontaneous localization of the position of one of the 10^{23} particles in the detection screen. We know that this single hit will immediately localize all of the other particles with which that one particle was entangled. So, one might argue that a localization in one of the three-dimensional subspaces of the wave function's space causes instantaneous action in many other of its three-dimensional subspaces. So there is action at a distance in the wave function's space. To be clear, this putative action at a distance wouldn't cor-respond in the low-dimensional image to the possibility of a meas-urement on A immediately affecting the spin of B. Rather it would correspond in the low-dimensional image to the possibility of a collapse of one of the atoms in the A-experiment's screen imme-diately collapsing all of the other atoms in that screen, as well as the spin state of A. All of this putative action at a distance in the high-dimensional metaphysics would correspond to action in the A-wing of the experiment in the low-dimensional metaphysics.

---

[10] Thanks to an anonymous reviewer for pressing this concern.

This is a more compelling concern than the former and presses us to think more carefully about the character of GRW hits in the high-dimensional space. In particular, it should prompt us to ask: are there distinct events that correspond to the hit in one of the three-dimensional subspaces and the hits in the other subspaces such that there is nonlocal action in the high-dimensional space? My answer to this question is: no. After the hit, we may view the wave function as having a certain peak at t. This single peak may be abstractly sliced up in different ways, as we view the wave function's space as being carved up into distinct subspaces. Such carvings may allow us to talk about "the initial GRW hit" and "the others." But there aren't in actuality distinct hits. There is just the one, though this one hit may be carved up in different ways. And so even if it is true that there is counterfactual dependence between what we may conceptualize as a peak in one subspace and a peak in the others, since we are not talking about distinct events when we talk about "the first peak" and "the others," there isn't causation on the table, as there would be if we were talking about some event taking place at one location in a space and another event taking place at another.

Perhaps an analogy will be of some use here. Consider a runner crossing a finish line to win a race. There is only one winning of the race and this is something the runner or the runner's body as a whole undergoes. We can, if we like, abstractly slice the runner up and think about just his surface's crossing the finish line. This then would let us ask questions about whether his surface's crossing the finish line caused the location of his other body parts at the moment he wins the race. But I think we should agree his surface's crossing the finish line is not a cause of his other body parts having the positions they do at the moment of his winning. The location of the runner's surface does have a special status if we compare it to the other abstract slices of the event: it is just the location of the runner's surface that is explanatorily relevant to his winning. Similarly, we can note, coming back to the case of interest, that there is just one abstract slice of the GRW hit that is explanatorily relevant to the hit's occurring. But the peaking of the wave function in this one of its three-dimensional subspaces is no more a cause of the other parts of the GRW hit than the location of the surface of the runner is a cause of the locations of the other parts of the runner. So again, there is no nonlocal action in the wave function metaphysics.

Finally, let us turn to no-collapse theories with additional variables such as Bohmian mechanics. Here, the wave function will behave the same as it does in the Everettian model. However, in this case, there will be some additional ontology, such as a particle (the so-called marvelous point) that moves around the wave function's space in a way described by the theory's guidance equation, that is, as determined by the state at a time of the wave function in the manner:

$$\frac{dQ_k}{dt} = \frac{\hbar}{m_k} \mathrm{Im} \frac{\psi^* \, \partial_k \psi}{\psi^* \psi} \left(Q_1, \ldots, Q_N\right).$$

The behavior of the marvelous point is determined by the state of the wave function in the neighborhood of the place in the high-dimensional space it occupies. And so again, there is no threat of nonlocal action. I have argued that Bohmian mechanics combines rather poorly with the interpretational framework of wave function realism. But if one is worried about fundamental nonlocality, a wave function realist interpretation of Bohmian mechanics could be of help.

Again we may ask, mirroring questions we asked in the section on separability, whether nonlocality is avoided only in the fundamental pictures of the world advocated by the wave function realist, or whether locality is preserved as well in the derivative metaphysics spread out in low-dimensional space or spacetime. The answer to this question again turns on whether the wave function ontology includes low-dimensional micro- and macroscopic objects as derivative realities.

Suppose it does.[11] Then one response the wave function realist may give to the question of whether there is nonlocal action in the derivative low-dimensional framework is that no, in that framework, there are correlations between spacelike separated events, but no genuine causal interaction because all such correlations have deeper explanations in terms of the behavior of the wave function. The more fundamental explanation for these correlations thus undercuts any causal explanation that may be provided by the existence of spatially distant events such as measurements on one half of an entangled pair.

I tried out such a line of reasoning in an earlier paper (Ney 2020a) as do Jenann Ismael and Jonathan Schaffer (2016). However, I now find such a position unsatisfactory.[12] The reason is that if one wants to argue in this way that there is no immediate causation across spatial distances because such causal relations are undercut or screened off by the behavior of the wave function, then one must similarly do so for all other causal relations in the low-dimensional framework. For there will always be a wave function explanation available at the more fundamental level. So, unless we are to be causal nihilists about what happens in the derivative low-dimensional space or spacetime, we should not argue that the behavior of the wave function undercuts the reality of derivative nonlocal action. What the wave function realist can offer is a more fundamental explanation of in virtue of what that derivative nonlocal action obtains, one that may give a more satisfying picture of what makes things happen in our world than one that contains unexplained nonlocal action. But it does not remove low-dimensional facts about nonlocal action.

Let's pause for a moment and see where we have arrived. I argued in Chapter 2 that wave function realism is not forced on

---

[11] If it doesn't, then there is no low-dimensional metaphysics to worry about whether it is local or not.

[12] Thank you to Jessica Wilson for pressing me on this issue.

us as a necessary consequence of the success of quantum theories. However, if one wants to take on the project of interpreting quantum theories as describing a world independent of our own minds, then wave function realism provides one a way of doing so. This is a way that reveals what appears in the low-dimensional representation of our ordinary perceptual experience to be objects spread out in space bearing irreducible relations and acting immediately across distances to be manifestations of a more unified, deeper reality, one which is separable and local.

Philosophers of physics have used different analogies to help us make sense of how the low- (three- or four-) dimensional appearances can arise out of a deeper reality that is higher-dimensional. Ismael once used the image of a kaleidoscope in which what appears in the viewing screen as a two-dimensional image of similar shapes appearing repeatedly are really the result of a smaller number of objects (beads, crystals) that are higher- (three-) dimensional (see Figure 3.3).

And Peter Lewis (2013) draws an analogy with the situation of the two-dimensional creatures in Edwin A. Abbott's story *Flatland*, beings confined to a plane in a larger, three-dimensional world. What appear to these creatures as things emerging out of nothing are lower-dimensional manifestations of objects moving

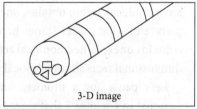

3-D image

One universe in a higher-dimensional space looks like many correlated objects in a lower-dimensional space

2-D image

Figure 3.3. The Kaleidoscope: Low- and High-Dimensional Images

through smooth trajectories in higher dimensions. Just as what is mysterious to the Flatlanders is unsurprising from the three-dimensional perspective, the wave function metaphysics reveals what is puzzling in the low-dimensional image to be an explicable consequence of wave-like behavior in a more fundamental high-dimensional one.

3.8 Avoiding Nonlocality with Nonseparability

We have now seen how the wave function realist avoids fundamental nonlocality. However, wave function realism is not the only strategy that has been pursued for providing quantum mechanics with a local metaphysics. Howard has argued that we can ensure a local metaphysics for quantum mechanics by rejecting the separability of quantum states. Indeed this is the central case he makes for the nonseparable metaphysics he favors, the quantum holism described in Chapter 2. He writes:

> If two systems are not separable, then there can be no interaction between them, because they are not really *two* systems at all. (1985, p. 173)

And, in light of the experimental confirmation of Bell's results,

> we must give up either separability or locality. . . . But if these are our only alternatives, then most of us would likely prefer the former alternative, on the grounds that special relativistic locality constraints are too much a part of our physics to be sacrificed to the cause of saving separability, all the more so because we have ready at hand a highly successful non-separable quantum mechanics. . . . In fact, I believe that Einstein himself would have followed us in this choice had he been forced to choose between these two alternatives. (1985, p. 197)

Howard's view is that when one accepts nonseparability, one thereby rejects the numerical distinctness of the individuals in entangled states. Turning back to the EPRB example, the proposal is that we reject the belief that the atoms are distinct entities and instead view them as one. Then we can explain the observed correlations without requiring nonlocality, without even requiring nonlocality in the low-dimensional image. There is no instantaneous influence of atom A on atom B because there is no distinction between atom A and atom B.

One may ask why we would be entitled to conclude just by the assumption that A and B are numerically identical that there cannot be causal influence between distant wings of the experiment. Surely a state of one object can cause another state of that self-same object. My being thirsty can cause my getting up to get a glass of water. What is more plausible, however, is that causal relations are irreflexive, that one state of an individual cannot cause that very state of the self-same individual. My being thirsty at a certain time t cannot cause my being thirsty at that same time t.[13] And so what Howard seems to be assuming is not or not merely that the atoms are numerically identical, but that the relevant states are identical (A's being measured x-spin up is identical with B's being measured x-spin down).

Strictly speaking, if it is really, as Howard notes, the rejection of nonseparability that allows one to avoid the conclusion of nonlocality, then perhaps we can avoid nonlocality using a less radical approach. After all, the rejection of numerical distinctness of states goes beyond what nonseparability requires. We can see this by considering the alternative nonseparable approach advocated by Teller and also the ontic structural realists. These take cases of quantum entanglement like the EPRB set-up we have been considering to involve what Teller calls "inherent relations," relations

---

[13] That causal relations are irreflexive is a standard assumption in the causation and causal modeling literature, e.g., Pearl (2000).

whose instantiation are not determined by intrinsic features of their relata.

But it may not be immediately clear how this view avoids nonlocality. According to relational holism, Bohm's atoms are numerically distinct, as are the states into which they enter on the two wings of the experiment. Thus it seems we have not done away with the consequence that a measurement on one atom instantaneously affects the result of a measurement on another some distance away. However, in "Relativity, Relational Holism, and the Bell Inequalities," Teller argues how it is supposed to be that relational holism avoids nonlocality. Teller grants that there will be stable correlations between the states of the two atoms in Bohm's set-up but argues that because of the inherent relations linking them, there is no reason to infer from these correlations to a causal mechanism linking the two wings. Because of the entanglement, the correlation may be brute:

> To say that causal locality has been violated most plausibly should be taken to mean that there are nonrelational properties of space-time points which are related in some other way—by action (lawlike dependencies) at a distance or through superluminal causal chains. On the other hand, when we are concerned with nonsupervening [or ungrounded] relations, this circle of ideas has no grip. There is no question of superluminal or distant action between nonrelational, definite values. (1989, p. 215)

and later:

> The correlation—as an objective property of the pair of objects taken together—is simply a fact about the pair. This fact will arise from and give rise to other facts. But it need not itself be decomposable in terms of or supervenient upon some more basic, nonrelational facts. There need be no mechanism into which the correlation can be analysed. (1989, p. 222)

Thus, Teller argues relational holism allows the Bell correlations to be brute, not requiring further explanation as the consequence of a causal relation between the distant events. It should be noted that as a sociological fact, ontic structural realists typically do not take on the project of using their nonseparability to avoid nonlocality. However if one liked Teller's approach, this would be a strategy available to them as well.

But it is not clear to me why Teller thinks that the pull to explain correlations is removed once we allow there are relations that are not determined by intrinsic features of their relata. The idea seems to be that once one gives up the assumption that *all* relational features are determined by intrinsic features of their relata (in other words, once one adopts relational holism), one should thereby give up the general presupposition that correlations have explanations. Indeed, he states that the adoption of relational holism frees us generally from all common cause reasoning. But even if relational holism makes it reasonable to allow *some* brute relations, it is not clear why it should remove the general presumption that correlations not be brute. To do so would seem to throw the baby out with the bathwater, giving up one of the most basic assumptions of scientific reasoning. Thus, I would argue that Howard's nonseparable metaphysics provides a more successful metaphysical motivation for the avoidance of nonlocality. Though, as I have noted, it involves more than a mere rejection of separability.

I have just raised concerns for Teller's strategy of using nonseparability to avoid nonlocality, but others have argued that even Howard's approach is insufficient to avoid the nonlocality that appears to arise in the case of the EPRB set-up. If this is so, then wave function realism would be an even more attractive option for those hoping to avoid nonlocality. So I will conclude this section by considering such a critique of Howard's approach.

In "Nonseparability Does Not Relieve the Problem of Bell's Theorem," Joe Henson argues that Howard is mistaken that one can

avoid nonlocality by adopting nonseparability even in the form of
holism. Howard had argued that it is generally assumed that the
systems on the two wings of an EPRB set-up are separable, numerically distinct systems. By rejecting that assumption, one can avoid
Bell's conclusion. But Henson argues that Bell's argument requires
only a weaker principle, that the systems on either wing of the experiment may be localized at distinct spatial regions, a principle he
calls Localized Events:

> All events can be associated to regions of spacetime in a con-
> sistent manner. (2013, p. 1012)

Since the argument doesn't require the stronger separability principle, one cannot avoid the conclusion (nonlocality) by rejecting it.
As Henson puts it:

> If one wants to rely on one's favourite derivation of Bell's theorem
> for the purposes of this discussion, one needs to show that the
> assumptions one makes are equivalent to, or weaker than, what is
> used (explicitly or implicitly) in the standard versions. After all,
> if I added the assumption that I live in London to a derivation of
> Bell's theorem, that would not make it reasonable for a group of
> angry realists to drive me out of town in the hope of saving lo-
> cality. (2013, p. 1009)

But does Howard's argument rest on this mistake?

As we have seen, Howard's main point is that the metaphysical
conclusion we should draw from experimental confirmations of
the quantum predictions is that the systems A and B are not numerically distinct. "If two systems are not separable, then there can be
no interaction between them, because they are not really *two* systems at all." And so although Bell may have not stated a separability
assumption explicitly in his paper, in describing the measurement
results as distinct events that may be localized at distinct locations,

we are thus assuming the negation of what Howard wants us to consider. So Howard could plausibly reject even Henson's weaker principle of Localized Events.

But it seems to me that nonseparability is a red herring here. By denying the distinctness of A and B and indeed the corresponding measurement events, Howard is denying that there even are two subsystems localized at distinct regions. Rather, there is just one system that isn't (at least straightforwardly) localized to one or the other region. So there is no issue of whether the facts about A and B determine the facts of the system contained at the union of the spacetime regions at which they are located. But whether what is at issue is nonseparability or not, there is a question about whether Howard's strategy is successful.

Henson argues that Howard's claim of numerical identity is unwarranted. He says:

> The suggestion . . . is that one could say to the worried experimenter "don't worry, when you saw the flashing light, that actually corresponded to an event in the whole experimental region, not an event in your lab. So you see, it's all local. . . . Actually your reaction to the flashing light didn't happen in your head either, but in the larger region." . . . It is in no way more unreasonable to apply this kind of reasoning to a hypothetical case of superluminal signaling than it is to apply it to outcomes in the EPRB experiment. (2013, p. 1020)

Henson's objection is essentially that Howard's rejection of numerical distinctness is ad hoc. I interpret him in this last sentence as pointing out that we could make a similar move in any hypothetical case of superluminal signaling and thus avoid nonlocality. But, he says, "if we rely on this, we may as well have avoided analysis of Bell's theorem by rejecting all locality assumptions except no-signaling in the first place" (2013, p. 1021).

I do not agree that the move Howard makes would be "no more unreasonable" to apply in any case of superluminal signaling. And this is because the denial of numerical distinctness in situations of quantum entanglement like the EPRB set-up is motivated independently of the desire to avoid nonlocality (or superluminal signaling). It is motivated also by the fact that in a three-dimensional ontology, it is simply not possible to give a complete account of what measurement results we should expect for A without considering the entangled system of which A appears to be only a part (and similarly for B). The atoms thus do not appear to have distinct realities.[14] This supports Howard's interpretative strategy.

But ultimately I am sympathetic with Henson's skepticism. Although Howard may avoid what is strictly speaking nonlocal interaction between distinct objects or events, I don't think his strategy to avoid nonlocality is as satisfactory as the wave function realist's. For recall that according to the wave function realist, the entire three-dimensional framework is derivative. Influence is transmitted instantaneously across spatial distances, but the source of these correlations is really a local ontology in a higher-dimensional space. On the other hand, Howard wants us to view reality as three-dimensional. We may say that what appear to be two distinct measurement events are not fundamentally distinct, but Howard will not deny that the single object and event is indeed spread out somehow in the three-dimensional space. And so although we might not have a causal interaction between two things, but only a state? or a process? involving only one, what we are committed to is still something that will involve a mysterious coordination across two distant parts of space at a time, a coordination that lacks explanation.

---

[14] The argument to this effect, not in any way relying on a desire to evade superluminal influence, was presented in Chapter 2. The motivation is a desire to capture the distinction between entangled and non-entangled states.

3.9 Motivating a Separable and Local Metaphysics

Assuming the wave function realist can provide a separable and local interpretation of at least some versions of quantum theories, in at least some sense of "local," we may now finally ask, why should we care? I will start with some empirical considerations and move toward others that are less empirical and more a priori.

I have already noted that wave function realism is not successful at securing the conceptions of locality used by Bell. And yet these are the senses of locality that some would say are mainly at issue when one worries about the incompatibility of relativity and quantum mechanics. It is generally thought that one of the main lessons of special relativity is that influences cannot be transmitted faster than the speed of light. And so one naturally wonders how quantum theories may be revealing a world in which objects affect each other instantaneously across (low-dimensional) spatial distances. There are legitimate concerns here and so we should begin by considering them. But perhaps the fact that wave function realism provides an ontology that fails to be nonlocal in its own space may help alleviate some of the concerns arising about nonlocality in four-dimensional spacetime.

In his recent book *Quantum Ontology*, Peter Lewis states the reason that nonlocal action is in tension with relativity in the following way. Suppose one allows that at least some of the time, objects can affect each other instantaneously across spatial distances. For example, something happening right here right now depends on the simultaneous mass of a distant star. According to special relativity, there are no absolute facts about which spatially distant events are simultaneous with which others. Spatiotemporal intervals are absolute, but facts about spatial distances or temporal durations always obtain only relative to a frame of reference or an observer's state of motion. So, there is no absolute fact about the mass of the star right now; what counts as right now is relative to an individual's reference frame. Lewis infers as a result that what this action at a distance is then is ill-defined according to relativity. Thus, it would seem, according to relativity, there cannot be action at a distance.

But although this way of understanding action at a distance as immediate influence between simultaneous events may be problematic from the perspective of special relativity, we can precisely state a related notion, defined in terms of the relativistic concept of spacelike separation. It after all does appear that Bell's results require us to believe there are, in cases of entanglement, influences between spacelike separated events.

There is division on the question of whether the quantum correlations we observe in the low-dimensional image are evidence of causal influences of this kind. Some have argued that cases of quantum entanglement do not require causal influence between spacelike separated events or the possibility of superluminal signaling (Maudlin 1994). More radically, the ER = EPR conjecture of Juan Maldacena and Leonard Susskind (2013) suggests that influence between spacelike separated entangled pairs may be mediated in a way compatible with relativity, through wormholes (Einstein-Rosen [ER] bridges). At the same time, however, others use contemporary frameworks for causal modeling, such as interventionist frameworks, to show how quantum entanglement does underwrite genuine causal influence (e.g., Shrapnel 2019) regardless of the existence of such wormholes.

Earlier I noted that one should not use the existence of an underlying wave function metaphysics to motivate a blanket denial of the existence of spacelike, that is, superluminal influences, unless wave function realists want to be wholesale causal nihilists about influences in spacetime. However, what looks puzzling from the perspective of a nonfundamental metaphysics may yet be revealed as explainable in terms of a more fundamental metaphysics. To the extent that wave function realism supports a derivative, low-dimensional ontology, it will yield an account of which spatiotemporal configurations exist at any given time in that derivative ontology. That will allow us to at least understand what is generating these conflicts with relativity, as it has often been understood, as barring superluminal influence.

In his 1948 paper "Quantum Mechanics and Reality," Einstein provided further reasons in favor of interpreting quantum theories in ways that achieved locality, and also separability. In both cases, Einstein used a kind of inductive argument, noting that the development of successful physical theories required the assumptions both of locality and separability.

Starting with separability, Einstein proposes:

> It appears to be essential for this arrangement of the things introduced in physics that, at a specific time, these things claim an existence independent of one another, insofar as these things "lie in different parts of space". Without such an assumption of the mutually independent existence (the "being-thus") of spatially distant things, an assumption which originates in everyday thought, physical thought in the sense familiar to us would not be possible. Nor does one see how physical laws could be formulated and tested without such a clean separation. (148, p. 321)

This passage points to what has been historically successful in the formulation of physical theories. They (a) give descriptions of isolated physical systems in terms of the arrangement of distinct objects or values at distinct locations and (b) predict and test what these physical systems will do from one time to the next based on how localized, individual systems are expected and observed to behave.

Einstein also argued that the assumption of locality seems to be required for the development of empirically successful physical theories:

> For the relative independence of spatially distant things (A and B), this idea is characteristic: an external influence on A has no

immediate effect on B; this is known as the "principle of local action," which is applied consistently only in field theory. The complete suspension of this basic principle would make impossible the idea of the existence of (quasi-) closed systems and, thereby, the establishment of empirically testable laws in the sense familiar to us. (1948, pp. 321–322)

The point seems straightforward enough. If what is nearby and observable may be affected by objects that are spatially distant, then without full knowledge of the occupants of the total spacetime manifold, how are we to make predictions about how the objects we observe will behave? Locality appears required to allow us to formulate testable empirical theories.

Now this point of Einstein's is itself contestable. In conversation, Wayne Myrvold has questioned it, noting that even in classical physics we are very comfortable writing down and testing laws knowing full well that there are spatially distant objects affecting the behavior of local ones. His example is the astrophysicist's description of the motion of Jupiter's moons. The Sun being 480 million miles away, Einstein's reasoning would lead one to believe that the physicist would need to reject its influence, modeling the behavior of the moon solely in terms of nearby factors. But this would produce wildly wrong results. This is a clear case in which the assumption of spatially distant influences is essential, not an obstacle to the formulation and testing of physical laws. Now of course what Einstein rejects is that an external influence on A has *immediate* effect on B, and one might respond to Myrvold by arguing that relativistic modeling will reject that the Sun's influence is immediate. But, astronomical phenomena are modeled quite well by Newtonian physics according to which gravitational influence is unmediated and instantaneous. Einstein thus seems wrong that physics simply cannot be done when we assume there are nonlocal influences and build these into our models.

Myrvold is right to object to hyperbole in Einstein's defense of locality in physics, but I don't believe this undermines a weaker defense of locality as an assumption guiding the formulation of tractable and testable physical theories. For in the case of the Sun and Jupiter's moons, the physics works because we are considering the influence of just a few large bodies a distance away. Things would devolve quite quickly if the modeling of Jupiter's moons needed to take into account immediate and significant influences from many or all distant bodies. So perhaps this is what Einstein is concerned about, thinking of widespread effects from quantum entanglement would massively complicate physics, perhaps leading to intractability. And so, for physics "in the sense familiar to us" to work, we discount immediate influence *in general and for the most part* where this is justified. Note that with advances in computational modeling, physics can take into account a much larger number of distant bodies successfully. However, this doesn't undermine Einstein's point, as in the age of big data, we are moving away from physics in the sense familiar to Einstein circa 1948.

So I propose we can make use of this weak point: we have good inductive reason to believe that physics of the kind that is already familiar to us, which involves modeling systems based on the assumption of separability and (mostly) local influence, is a way of developing inductively successful theories. But can this justification for a separable and local metaphysics be used to generate any support for wave function realism? I am afraid it cannot. The point is, after all, that the testing of laws depends on our ability to manipulate and observe what is happening at a confined region of space, isolating objects from outside influences. But of course, the space in which human beings' manipulations and observations take place is not the high-dimensional space of the wave function. Thus, it seems, Einstein's defense of separability and locality justifies a separable and local ontology in three-dimensional space or spacetime, the framework in which we interact with objects, but not a separable and local wave function ontology.

Perhaps another case to be made for separable and local interpretations of physical theories may be found in the work of Valia Allori. Allori (2013a) defends another view she finds in Einstein, that "the whole of science is nothing more than a refinement of our everyday thinking." She elaborates:

> The scientific image typically starts close to the manifest image, gradually departing from it if not successful to adequately reproduce the experimental findings. The scientific image is not necessarily close to the manifest image, because with gradual departure after gradual departure we can get pretty far away. . . . The point, though, is that the scientist will typically tend to make minimal and not very radical changes to a previously accepted theoretical framework. (2013a, p. 61)

One might then say since our pre-scientific thinking and subsequent physical theories postulate separable and local ontologies, that our quantum theories should, where possible, do so as well.

To be clear, Allori is herself not making this point to argue for the separable and local ontology of wave function realism. She is using the point to argue for her preferred primitive ontology framework, since she believes that all previous (i.e., nonquantum) physical theories also possessed a primitive ontology.[15] But one might hope that her point extends to make a case for a wave function ontology as well.

Unfortunately, I don't think it does. Because wave function realists also reject as fundamental a three-dimensional spatial background, replacing it with an unfamiliar, high-dimensional background, it is not really so plausible to argue that *this* separable and

---

[15] For an argument that wave function realism should be rejected for its radical departure from the manifest image, see also Emery (2017).

local ontology is closer to the manifest image and classical theories than one that would jettison one or both of separability and locality, but retain the low-dimensional spatial background of our experience. If we agree with Allori that minimal departures should, where possible, be preferred, the move to higher dimensions is very far from a minimal departure.

Let's now move to consider more purely a priori reasons in support of separable and local ontologies to see if any of these may earn some support for the metaphysics of wave function realism. Some of these were brought to bear in the eighteenth century as natural philosophers struggled with Newton's characterization of gravitational forces as acting immediately across spatial distances. Newton himself sometimes claimed that action at a distance is impossible— for example:

> The cause of gravity is what I do not pretend to know and therefore would take more time to consider of it. . . . That gravity should be innate, inherent, and essential to matter, so that one body may act upon another at a distance through a vacuum, without the mediation of anything else, by and through which their action and force may be conveyed from one to another, is to me so great an absurdity that I believe no man who has in philosophical matters a competent faculty of thinking can ever fall into it. (Newton 2007)

To claim nonlocality is absurd is not thereby to offer an argument against it. Nor to my knowledge did Newton ever offer a clear argument for why action at a distance is absurd; however, we do find something in the work of Newton's disciple Clarke, in his correspondence with Leibniz:

> That one body should attract another without any intermediate means, is not a miracle, but a contradiction: for 'tis supposing

something to act where it is not. But the means by which two bodies attract each other, may be invisible and intangible, and of a different nature from mechanism; and yet, acting regularly and constantly, may well be called natural. (Ariew 2000, p. 35)

Clarke, like Newton, supposes that gravity must act locally even if the means by which it does so may be invisible. And the reason this must be so is for something to act, it must be located where it acts. Otherwise, it wouldn't be it itself that is so acting, but something else, or nothing at all. There is something, I believe, that is sensible about this point and it explains at least one reason that nonlocal action strikes us as deeply unintuitive and worse, incoherent. And it is a consideration that may be brought to bear in support of wave function realism's local ontology.

What about separability? There is something intuitively compelling as well about the idea of separability, that there (i) exists a basic class of independent objects arranged at distinct spatial locations and that (ii) what things are like at any composite region is ultimately determined by the features of these more basic objects. Teller, as we have noted, questions the coherence of the holist's rejection of (i), the claim that what appears to be a multiplicity of things spread out in space are not distinct after all (1986, p. 73). It is unusual, however, for philosophers to question the very coherence of rejecting (ii).

That one should endorse (ii) is not so much enforced as a matter of logical coherence, but rather as a matter of clarity. It is embraced in David Lewis's metaphysical doctrine of Humean supervenience, presented by him in the following way:

Humean supervenience is named in honor of the great denier of necessary connections. It is the doctrine that all there is to the world is a vast mosaic of local matters of particular fact, just one little thing and then another. . . . We have geometry: a system of external relations of spatiotemporal distances between points. . . . And at those points we have local qualities: perfectly natural intrinsic properties which need nothing bigger than a point at which to be instantiated. For short: we have an arrangement of qualities. And that is all. There is no difference without difference in the arrangement of qualities. All else supervenes on that. (1986a, pp. ix–x)

Lewis recognizes Humean supervenience as a contingent doctrine. It could be the case that there are further relational facts, not determined by the arrangement of qualities. As he puts it, "It is not, alas, unintelligible that there might be suchlike rubbish" (1986a, p. x). But he proceeds with the assumption that "suchlike rubbish" does not exist. Ironically, Lewis defended Humean supervenience because he thought this was the way physics represented the world (1986a, pp. x–xi): "If Humean supervenience is true at all, it is true in more or less the way that present physics would suggest." As we are seeing in this book, it is actually rather contentious what present physics suggests about Humean supervenience and separability. Many think that quantum theories should force us to question separability (and so Humean supervenience).[16] And so we cannot use quantum mechanics to bolster these doctrines. Instead, we must ask first what can be said for separability on its own terms.

If separability is not logically or metaphysically necessary, it is at least intuitive in the respect of being simple. As Daniel Nolan remarks in a discussion of Humean supervenience:

Humean supervenience is in some respects a very parsimonious theory. Not only does it not have any fundamental relations besides spatiotemporal ones, or any fundamental properties of

---

[16] See, e.g., Maudlin (2007b) for a critique of Humean supervenience based on an argument that quantum entanglement should make us give up separability.

more than point-sized things, it also has no facts or truths that
are true even though they depend on nothing in particular. . . .
Perhaps we shall need to postulate things not covered by Humean
supervenience to achieve a proper explanation of all the phenomena in the world. But it would be good to get a clearer idea
of what requires something extra, and when we are obliged to go
further. (2005, p. 31)

It is just simpler to believe in a basic set of entities each with their
own locations upon which everything else depends. This makes a
separable metaphysics a promising starting point for inquiry. And
this is how I read the argument in Loewer (1996) using Humean
supervenience to motivate wave function realism, that we do not
have to move to anything more complicated, postulating any additional relations or relational facts (Lewisean "rubbish") to adequately account for the reality of quantum entanglement.

3.10 In Defense of Intuitions

It is my view that the best case the wave function realist has for developing her distinctive ontological framework comes from such
conceptual considerations and intuitions.[17] But one might question
whether it is at all desirable to have an interpretation of quantum
theories that conforms to our intuitions. Ladyman and Ross (2007)
criticize such interpretational projects, calling them domestications
of science. But it is my attitude that quantum theories stand very
much in need of domestication to the scientific community and

[17] And so it may be for some of the rival frameworks as well. For we have already noted
Allori's appeal to conform one's metaphysics to "everyday thinking" in support of the
primitive ontology framework. The issue of which framework to move forward with in
some cases may thus boil down to the issue of which intuitions one takes to be more important or useful for the project at hand.

greater public.[18] The discovery and development of interpretations that are compatible with our intuitions may be useful for a number of reasons. I will mention three such benefits that conceptually clear and intuitive interpretations may bring. All are unabashedly pragmatic.

First, an ontological interpretation of a physical theory, by providing one with a clear account of what the world is like according to that theory benefits students and scientists in allowing them a clearer handle on the theory they are working with. Although it is not possible to understand our best scientific theories without having a handle on the mathematics used to state them, a clear metaphysics to supplement the mathematics can be instrumental in seeing more clearly what a given theory says, allowing one to more easily learn and use it. As an example, the special theory of relativity before it is supplemented with the clear interpretation of a four-dimensional Minkowski spacetime manifold can seem to lead to paradoxes in measurements that are difficult to comprehend: the paradox of the train and the tunnel, the twins paradox. These are not genuine paradoxes; there is no such inconsistency in the theory. But comprehending this fact is much easier when the mathematics is supplemented with a picture of entities spread out in four-dimensional spacetime, for which facts about elapsed time or spatial distance fail to be absolute. I believe something similar can come to pass for quantum theories. Once supplemented with a clear metaphysics, what looks paradoxical or surprising becomes clear, natural, and easy to work with. And there is no reason distinct interpretations cannot produce alternative accounts useful in this respect.

Second, an ontological interpretation says things that go beyond what the theory on its own says and in this respect, such interpretations can be fruitful in generating new speculations or

---

[18] This is not to deny that the project of domestication has already been carried out to a large extent by the work of those providing clear solutions to the measurement problem.

predictions that can then extend the theoretical power of the theory. Should one adopt the wave function metaphysics and its attendant higher dimensions, one can begin to ask more questions about the structure and contents of these higher-dimensional spaces and learn more things about them that would simply not be discussed without attention to this question of ontology.

Finally, for myself and many other former physics students, the reason we chose physics as a focus of study in college was to learn about the fundamental nature of reality. Without an ontological interpretation, physics doesn't provide this. Under the influence of Copenhagen, Mermin's "Shut up and calculate!" and Feynman's "I think I can safely say no one understands quantum mechanics," students often come to quantum theories puzzled about what they say about the world, but then are told not to ask such questions because the theory is impossible to understand. This is disappointing and drives students out of the field. Not all physics students care about questions of ontology and the deep issue of the nature of reality, but for those who do, it is worth having serious work on ontological interpretation that can give them what they are looking for. We need more, not fewer students of physics.

I don't want to leave the reader with the sense that anything I am saying challenges the idea that we should not at the same time work on ontological interpretations that challenge our thinking. In fact, all of the interpretations of quantum theories that are available have aspects of unintuitiveness—this is simply unavoidable in the interpretation of quantum theories. And it is what is so exhilarating about the study of these theories, how they challenge what we previously thought was obvious. What is being suggested in this last section, however, is that there is nothing problematic about trying to fit these startling aspects of the world into a picture we can understand.

The wave function realist need not deny that there are clear senses of separability and locality for which there are nonseparable facts and nonlocal influences. The question is whether one should take these to be brute facts about our world or attempt to provide explanations in terms of an underlying metaphysics. Wave function realism is such an attempt at explanation. The virtues of having interpretative options that provide such explanations justify the exploration and development of this framework, and exhibit one respect in which it is superior to the others, and may very well be correct.

4

Wave Function Realism in
a Relativistic Setting

4.1 Removing Idealization

As we saw, there is a prima facie case for wave function realism based on the ubiquity and success of wave function representations in quantum theories. Wave function realism is not the only way to provide an ontology that can support these representations and explain what kind of world these representations describe; but, as we saw in Chapter 3, the sorts of pictures of the world that wave function realism yields present us with some intuitively nice metaphysical features. And having interpretations of quantum theories that have such features may bring with them practical benefits.

Despite these virtues, an important class of criticisms has been raised to wave function realism, and these gain traction based on the fact that wave function realism has until now been formulated and defended solely within the context of idealized, nonrelativistic quantum mechanics. Although nonrelativistic quantum theories are useful as approximations, teaching us important lessons about a range of real-world phenomena, it seems that if one is going to learn things about the fundamental structure of our world from a consideration of quantum theories, these are going to have to be lessons that remain even when we move beyond the nonrelativistic to the relativistic domain. After all, we know our world to be a relativistic world. With this in mind, Wayne Myrvold, Chris Timpson, and David Wallace have all argued that wave function realism yields pictures of our world that are reliant on features of quantum

The World in the Wave Function. Alyssa Ney, Oxford University Press (2021). © Oxford University Press.
DOI: 10.1093/oso/9780190097714.003.0004

theories that do not carry over to the relativistic setting (Wallace and Timpson 2010, Myrvold 2015, Wallace 2020). Therefore, wave function realism does not adequately capture aspects of the fundamental nature of our world, according to quantum theories.

It is an interesting question whether wave function realism must, to be viable as a framework for interpreting quantum theories, have application beyond the domain of nonrelativistic quantum mechanics. Must a framework for the ontological interpretation of a quantum theory be workable as an interpretation for all quantum theories? I do not see why it must. I believe that the development of wave function realism as an interpretation even only of nonrelativistic quantum mechanics could be useful for all of the reasons described in Chapter 3. Moreover, one might argue that solutions to the measurement problem have actually only been worked out clearly and adequately in the context of nonrelativistic quantum mechanics.[1] As we discussed in Chapter 1, the project of the ontological interpretation of physical theories begins with those that make contact with the world of our experience; therefore, it is not clear whether it is appropriate to be interpreting theories without a solution to the measurement problem.

Nonetheless, I am interested in what the application of a wave function realist framework to our best quantum theories—those that are the most general, least idealized, and best supported—may have to teach us about the fundamental nature of our world.[2] And so, in this chapter, we will take a look at how if, at all, wave function realism may be applied to relativistic quantum theories, including those quantum field theories making up the Standard Model of particle physics.

---

[1] Thanks to Valia Allori here.

[2] It can be questioned whether quantum field theories are more general than nonrelativistic quantum mechanics. This comes down to which kinds of theories are applicable to the most phenomena in nature. It may be argued that actually in physics, nonrelativistic quantum mechanics gets applied to more phenomena; however, I won't engage in this bean-counting exercise here.

4.2 Five Critiques

I count five different critiques of wave function realism that cite the difficulties of extending it to a relativistic setting. I will describe them in chronological order of publication. A later section will examine replies on behalf of the wave function realist.

A. The first critique, raised by Wallace and Timpson (2010), is that there is no good account of the sort of space the wave function is meant to be defined on when wave function realism is considered as an interpretation of relativistic quantum field theories. In the nonrelativistic case, recall, the wave function is defined on a space that has the structure of a classical configuration space. This is a space with $3 \times N$ dimensions, where N is the number of particles in the universe, particles that are claimed to be ontologically derivative. For the wave function realist, this provides a "top-down" characterization of the fundamental space of quantum mechanics, since particles are not fundamental; but it is one that nonetheless singles out a basic structure. For nonrelativistic quantum mechanics, the character of the configuration space is straightforward, since particle number is assumed to be conserved. However, in quantum field theories, particle number is not conserved. Particles may be created and destroyed. Moreover, systems may evolve into states describable as superpositions of particle number. This undermines the possibility of understanding the world as simply constituted by a relativistic wave function as a field on some unique, determinate configuration space. Wallace and Timpson consider an alternative possibility, namely, that for quantum field theories, the wave function realist should instead postulate an infinite number of (non-normalized) wave functions: a single-particle wave function living on a three-dimensional space, a two-particle wave function living on a six-dimensional space, and so on.[3] However, they (rightly) assume that the wave function realist will not prefer to adopt such an ontologically profligate metaphysics.

B. A second objection is that wave function realism "obscures the role of spacetime in quantum theories." This objection is also presented by Wallace and Timpson, who note that in all standard presentations of quantum field theories, field operators are assigned to localized spacetime regions. (This is true as well in algebraic quantum field theories, where algebras are associated with spacetime regions.) Systems are not described as in the Schrödinger wave-dynamical representation of nonrelativistic quantum theories in terms of the evolution of a wave function in configuration space. And so, the representations on which wave function realism is based simply are not standard representations in relativistic quantum theories.

C. A third critique relates to the demand of relativistic covariance. To get a theory of the evolution of the wave function that is relativistically covariant, one sacrifices what David Albert (2015) has called "narratability." That is, there will be no unique correct account of how the wave function evolves from one time to the next. Wallace and Timpson (2010) press this failure of narratability not as leading to a decisive refutation of wave function realism. Rather, they argue that this failure of narratability is obscure and surprising in the context of wave function realism, where one is supposed to be achieving a separable fundamental metaphysics. It is, by contrast, unsurprising in the context of rival frameworks for the interpretation of quantum theories that are not intended to

---

[3] This is an approach one finds implemented in Bohmian approaches to quantum field theory, e.g., Dürr et al. (2013), p. 241.

be separable, including the one they prefer, spacetime state realism.

D. A fourth critique aims to show that in the context of relativistic theories especially, the existence of a wave function is derivative on the antecedent existence of structures defined on ordinary spacetime. Myrvold (2015) shows how wave functions, to the extent that they may be constructed in quantum field theories, may be defined in terms of the global quantum state, the vacuum state, and field operators associated with points of spacetime. So, he concludes, the wave function cannot be more fundamental than a spatiotemporal ontology, as the wave function realist believes.

E. Finally, Wallace argues that the privileging of the position basis is problematic in the context of quantum field theories, for which quantum states and observables are more typically defined in terms of a momentum basis (Wallace 2020). We can see that at least in the nonrelativistic case, the wave function realist does privilege the position basis as she defines the wave function as an assignment of values to points in a space with the structure of a classical configuration space, as opposed to some other kind of state space. One might think it is unproblematic that quantum field theories generally characterize states in terms of a momentum basis, as one can simply Fourier transform to achieve a position representation, as described in Chapter 1. But this is not straightforward in the case of quantum field theories. There it turns out that even if representations of exact states of position can be achieved, these have the consequence of violating Lorentz covariance (cf. Teller 1995, pp. 85–90). And so, if the privileging of position is an essential feature of wave function realism, not merely confined to the interpretation of nonrelativistic quantum mechanics, then wave function realism cannot be applied to construct a viable interpretation of relativistic quantum theories.

These are the objections. What the three critics making these objections prefer is an interpretational framework according to which quantum theories describe an assignment of values (operators, algebras) to spacetime regions rather than to the high-dimensional space preferred by the wave function realist. This is the view we discussed in Chapter 2 that Wallace and Timpson label *spacetime state realism*. Adopting such a framework avoids the privileging of the position basis and the dependence on a fixed number of particles that the framing of an ontology in terms of a configuration space representation requires.

On the other hand, spacetime state realism lacks the separability of the wave function realist's metaphysics. And so one might ask whether the wave function realist might be able to supply a high-dimensional picture that is appropriate for quantum field theories, in failing to privilege the position basis and allowing for variation in total number of particles, while retaining fundamental separability and locality. For the motivations for the high-dimensional picture carry over to the relativistic case, even if the applicability of a high-dimensional *configuration space* representation does not. Shrugging off the constraints of the wave-function-in-configuration-space view, we can begin to see what a more plausible relativistic application of wave function realism may look like and how one may respond to the five criticisms listed above.

4.3 Wave Function Realism for Relativistic Quantum Theories

As argued in Chapter 3, the advantage wave function realism has over rival realist interpretations of quantum theories is that only wave function realism provides a metaphysics that is fundamentally both separable and local. In the nonrelativistic case, we developed this argument by considering what initially appear to be spatially separated particles of a determinate number. At least prima facie, the case for wave function realism may be extended by considering any type of event occurring at distinct spacetime locations for which there appear to be correlations induced by entanglement. The wave function realist will take these correlations to suggest that what appear to be distinct events occurring at distant spacetime locations are manifestations of a more fundamental separable and local ontology in a higher-dimensional framework.

To see how this extension of the argument of Chapter 3 may proceed, we start from the assumption that our quantum field theory represents systems using an assignment of operators to regions of spacetime. From this supposition, Wallace and Timpson infer that the ontology of these theories must involve features instantiated at spacetime regions. They elaborate their spacetime state realist ontology in the following way:

> Suppose one were to assume that the universe could be divided into subsystems. Assign to each subsystem a density operator. We then have a large number of bearers of properties—the subsystems [they take these to be spacetime regions]—and the density operator assigned to each [spacetime region] represents the intrinsic properties that each subsystem instantiates, just as the field value assigned to each spacetime point in electromagnetism, or the complex number assigned to each point in wavefunction realism, represented intrinsic properties. (2010, p. 709)

Wallace and Timpson also allow that one can think of the ontology in terms of an assignment of algebras of operators to each spacetime region rather than operators of some kind (2010, p. 712). Either way, their point remains that the fundamental ontology of quantum theories, including quantum field theories, involves an assignment of values of some kind to regions of spacetime, rather than an assignment of values to regions of some higher-dimensional space.

Wallace and Timpson both acknowledge and embrace the fact that although relativistic quantum systems can be characterized in terms of an assignment of algebras or operators to spacetime regions, interpreting this literally leads to a nonseparable metaphysics. Facts about the assignment of such features to the subregions of a spacetime region R do not determine the assignment of features to R.

We may debate the importance of having an ontology that is separable. Wallace and Timpson question the importance of separability; the previous chapter attempted to lay out the case for separability and locality in some detail. But like it or not, wave function realists are motivated by the desire to have a separable ontology, and there does not seem to be any prima facie argument that the desire for a relativistic theory conflicts with the desire for a picture of the world that is fundamentally separable and local. And so spacetime state realism, although it may have the virtue of producing an ontology that is close to the mathematical structure of quantum field theories (insofar as that is a virtue), will only be a stopping point for the wave function realist on the way to a more fundamentally separable ontology.

The simplest way for the wave function realist to proceed in developing an interpretation of relativistic quantum theories then would be in many respects analogous to the way she achieved the high-dimensional representation in the nonrelativistic case. Starting from an apparent ontology of localized subsystems in three-dimensional space with nonseparable features, the wave function realist posited a more fundamental, higher-dimensional space in which each point corresponded to an entire three-dimensional configuration of subsystems. And now we will see, in the relativistic case, starting from a nonseparable ontology of field operators assigned to localized spacetime regions, the wave function realist will posit a more fundamental higher-dimensional space in which each point corresponds to an entire four-dimensional configuration of field operator assignments.

The aim of relativistic quantum theories like quantum field theories is, not surprisingly, to achieve quantum theories that are

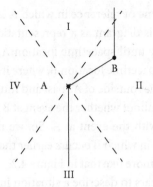

Figure 4.1. A Particle Trajectory

compatible with relativity, special relativity in particular. One consequence of special relativity is that temporal facts are not absolute and depend on a frame of reference. In particular, for spacelike separated events A and B, whether these events occur simultaneously, A occurs earlier than B, or B earlier than A depends on a frame of reference. It is interesting to see how this immediately leads to processes in which the total number of particles in a system may change or be indeterminate.[4]

To see this, consider the spacetime diagram in Figure 4.1. Dotted lines indicate the path light rays would take in and out of point A. Solid lines indicate particle trajectories (or worldlines). Quadrant I represents the future lightcone of A, that is, all spacetime locations a signal from A could reach traveling at a speed less than that of light. Quadrant III represents the past lightcone of A, that is, all spacetime locations from which a signal could reach A traveling at a speed less than that of light. Quadrants II and IV indicate all locations that are spacelike separated from location A. A signal coming from any one of the locations in quadrants II or IV would have to travel at a speed greater than that of light to reach A.

<hr>

[4] Thanks to Markus Luty here.

Assuming a frame of reference in which A is earlier than B, it is natural to view this diagram as a representation of a particle that remains stationary until spacetime location A, and then veers off to the right until spacetime location B, where it again comes to rest. However, since B lies outside of A's past and future lightcones, there is no absolute fact about whether the event at B is earlier, later than, or simultaneous with the event at A. So, we may also consider a frame of reference in which B occurs earlier than A. Then we could view the situation more like that in Figure 4.2.

This figure appears to describe a situation in which again a particle remains stationary until spacetime location A. However, from this frame of reference, it appears that an event at B creates two particles, one of which moves to the left, annihilating the first particle at A. The other continues on at rest. So what do we find at location A: a situation in which there is one particle, or a situation in which there are two? The answer is that both are legitimate particle-language descriptions of the situation. And, interestingly, relativity implies that, from the perspective of some frames, there are processes that create and destroy particles. Particle number may vary from one time to the next.

In relativistic quantum theories, the fact that particle number may vary is accommodated using the tool of a Fock space. In

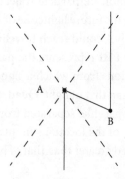

Figure 4.2. Particle Creation and Annihilation

Fock space, there is a vacuum state, denoted | 0>, which may be interpreted as a state in which there are no particles. There are also single particle states, denoted | p>, states in which there is a single particle with momentum p. And there are many particle states, denoted | p_1, p_2, ... p_n>, states in which there are n particles with momenta p_1, ..., p_n. Finally, following the principle of superposition described in Chapter 1, there are all linear combinations of the preceding kinds of states.[5] If there are different species of particles, for example, particles with different masses, such as ϕ-particles with mass m_ϕ, and ψ-particles with mass m_ψ, then these states can be denoted in this way as $|\phi(p_1), \ldots \phi(p_n), \psi(p_{n+1}), \ldots \psi(p_{n+m})>$.

In nonrelativistic quantum mechanics, the basic operators that act on states are the position and momentum operators. We can use these operators to define creation and annihilation operators, operators that create and destroy particles. The creation operator $\hat{a}^\dagger(p)$ creates particles with momentum p. The annihilation operator $\hat{a}(p)$ destroys particles with momentum p. So, $\hat{a}^\dagger(p)|0> = |p>$ and $\hat{a}(p_1)|p_1, p_2> = |p_2>$

Quantum field theory uses such Fock states and operators in order to build up an inventory of local observables.[6] It is these local observables that are taken to be essential in building a quantum field theory that can describe interactions in nature.[7] The interest in having local observables comes because the aim is to have a relativistic theory, where this is taken to require that there be no causal influence between spacelike separated events. In other words, we want to ensure local causality in Bell's sense.

---

[5] We can see we are already incorporating two of the points about quantum field theories raised in the critiques of wave function realism listed in Section 4.2: that particle number may vary and that states are typically represented using a momentum rather than position basis.

[6] The presentation here will follow that of the early chapters of Preskill (2019) on simple quantum field theories for scalar field interactions.

[7] If we want a theory that can solve the measurement problem, we need a theory that can accommodate measurements, cases of interactions.

The observables of a quantum theory are given by a certain class of operators. These are the self-adjoint or Hermitian operators.[8] The local observables are understood as those operators that commute at spacelike separation. In other words, for a given state ψ, where locations x and y are spacelike separated, the operators $\hat{O}(x)$ and $\hat{O}(y)$ commute if it does not matter if we operate first on ψ with $\hat{O}(x)$ and then with $\hat{O}(y)$, or if we operate first on ψ with $\hat{O}(y)$ and then with $\hat{O}(x)$; either way, we get the same result. This ensures local causality by requiring essentially that causal tinkerings at spacelike separation from some system have no immediate effect on that system.[9]

It is not trivial to construct local observables in a given quantum field theory, but using what we have built up to this point, we may just state the local observables that are used in the simplest quantum field theories to describe the states of bosonic scalar fields, what would be used to describe systems of scalar (spinless) particles without charge like the Higgs boson.[10] The simplest such operator is:

$$\hat{\phi}(x) = \int \frac{d^3p}{(2\pi)^{3/2}\sqrt{2\omega_p}} [\hat{a}(p)e^{-ip\cdot x} + \hat{a}^{\dagger}(p)e^{ip\cdot x}]$$

It does indeed turn out that for any state ψ, $[\hat{\phi}(x)\hat{\phi}(y) - \hat{\phi}(y)\hat{\phi}(x)]\psi = 0$. So, $\hat{\phi}(x)$ is a local observable. This is the basic field from which one can construct other observables in a scalar field theory. In particular, the interaction Hamiltonian

---

[8] See, e.g., Griffiths (2005), p. 100 or Shankar (2012), p. 27. These two notions are generally taken as equivalent. An operator is self-adjoint or Hermitian if it is identical to its transposed complex conjugate.

[9] I'm focusing on the role of local observables in conventional quantum field theory here, but see Ruetsche (2011), pp. 106–107, for a discussion in the context of algebraic quantum field theory.

[10] The consideration of quantum field theories for fermionic particles, or those with charge or spin, would not affect the general ontological points that follow.

of a quantum theory, what describes a system's evolution under interactions, will be constructed from such an observable.

The standard way to compute the probabilities of transitions or interactions in quantum field theory is S-matrix theory. S-matrix theory is sometimes criticized from a foundational point of view, first since it relies on what are clearly idealizations, and second because of questions about renormalization.[11] However, it is the predictions generated by S-matrix theory that have led to the enormous empirical successes of the Standard Model of particle physics, and so these criticisms are now standardly considered to be misplaced at least insofar as they suggest we should not trust S-matrix theory to give us a good model of interactions at the relatively high energy scale at which we test the Standard Model. It is a separate question whether one should trust the method for all energy scales.

S-matrix theory tells us to consider the initial and final states of a causal process to consist of noninteracting particles at distinct locations. We then evolve the initial state forward using an interaction time evolution operator (the S-matrix) which gives us the probability of the process occurring as: | <final state| $\hat{S}$ | initial state>|^2. The following example will be helpful in answering questions about the ontology of quantum field theories in a wave function realist framework, and so I will present the outline of how transition probabilities are arrived at for this case.

Suppose we have a theory with two scalar fields, $\hat{\phi}(x)$ and $\hat{\psi}(x)$, where the former as we saw is built out of the creation and annihilation operators $\hat{a}^\dagger(p)$ and $\hat{a}(p)$:

$$\hat{\phi}(x) = \int \frac{d^3p}{(2\pi)^{3/2}\sqrt{2\omega_p}} \left[\hat{a}(p)e^{-ip\cdot x} + \hat{a}^\dagger(p)e^{ip\cdot x}\right]$$

[11] See Teller (1995) for illuminating discussion of both issues, and Fraser (2020) for discussion of some others.

The latter is built out of creation and annihilation operators $\hat{b}^\dagger(p)$ and $\hat{b}(p)$:

$$\hat{\psi}(x) = \int \frac{d^3p}{(2\pi)^{3/2}\sqrt{2\omega_p}} \left[\hat{b}(p)e^{-ip\cdot x} + \hat{b}^\dagger(p)e^{ip\cdot x}\right]$$

Suppose we want to evaluate the probability of the following interaction taking place, an interaction in which two ψ particles with initial momenta p_1 and p_2 scatter off each other and leave with momenta k_1 and k_2. The initial state of the system may be regarded as a state of noninteracting particles written as $|\psi(p_1), \psi(p_2)>$. The final state is $|\psi(k_1), \psi(k_2)>$. What we want to compute is $\left|<\psi(k_1), \psi(k_2)|\hat{S}|\psi(p_1), \psi(p_2)>\right|^2$. The S-matrix depends on the correct theory of interactions here, the right interaction Hamiltonian we should use, but may be summarized as the following.[12]

$$\hat{S} = Te^{-i\int_{-\infty}^{\infty} dt\hat{H}_{int}(t)},$$

We can suppose the interaction Hamiltonian depends on the fields and not their derivatives, in which case this may be cast in terms of an interaction Lagrangian as:

$$\hat{S} = Te^{i\int_{-\infty}^{\infty} d^4xL_{int}(\hat{\phi}(x))},$$

We will work with $L_{int} = \frac{1}{2}g\hat{\phi}\,\hat{\psi}\,\hat{\psi}$, where g is a number. This is the so-called interaction vertex for our theory and g the strength of the interaction. It provides the basic rules for what the interactions

---

[12] The 'T' indicates this is a time-ordered exponential.

Figure 4.3. Interaction Vertex

look like. At each vertex, or spacetime point in this case, there is one $\hat{\phi}$ field and two $\hat{\psi}$ fields (Figure 4.3). More complicated theories will have more varieties of interaction vertices. The dashed line represents a $\hat{\phi}$ field, the solid lines, $\hat{\psi}$ fields.

After we expand the exponential into a power series up to the first nonzero term, we will arrive at the following integral:

$$\frac{i^2}{2!}\left(\frac{g}{2}\right)^2 \int d^4x_1 d^4x_2 <\psi(k_1), \psi(k_2) \left| T(\hat{\phi}\hat{\psi}\hat{\psi})(x_1)(\hat{\phi}\hat{\psi}\hat{\psi})(x_2) \right.$$
$$\left| \psi(p_1), \psi(p_2)> \right)$$

I won't work through the calculation of this integral here. The important upshot for us, however, is that it leaves us with a sum of three terms that can each be summarized in the form of a Feynman diagram. In quantum field theory, one learns how to arrive at these diagrams quickly and uses them to directly calculate scattering amplitudes, and hence transition probabilities. The sum in this case may be written as in Figure 4.4, with each term weighted by a symmetry factor we will not calculate here.

We now have enough formal machinery to be able to address questions about the ontology of quantum field theories. Taking a look at these diagrams, what is being suggested is that a system undergoing a $\psi\psi \rightarrow \psi\psi$ interaction is in a superposition of three different processes. In the first, the initial ψ particles annihilate each other at the first location x_1 to create a ϕ particle that is transmitted to x_2 and then creates two new ψ particles. In the

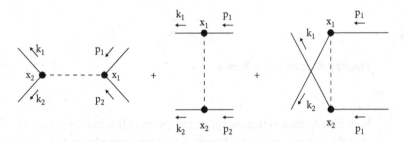

Figure 4.4. Feynman Diagrams

second, one ψ particle arrives at x_1, the other arrives at x_2, and a single ϕ particle mediates their change in momenta (from p_1 to k_1 and from p_2 to k_2). In the third, the ϕ mediates a distinct change in momentum (from p_1 to k_2 and from p_2 to k_1). These are different channels for the interaction, that is, different intermediate interactions we might observe, were we to probe the system appropriately. In the context of a hidden variables approach to the measurement problem like Bohmian mechanics, we may view them as indicating different paths our ψ- and ϕ-particles might actually take. But in the context of a non-hidden variables approach, we should instead view the system as in an entangled state, with different amplitudes applying to different total spacetime scenarios.

In this way, we can see the nonseparability in the (low-dimensional) spacetime representation of the interaction. For example, if there are particles with momenta p_1 and p_2 at location x_1, then there are particles with momenta k_1 and k_2 at x_2. And if there are particles with momenta k_1 and p_1 at x_1, then there are particles with momenta k_2 and p_2 at x_2. Just as in the EPRB case, where entanglement brought with it irreducible relations between the spins instantiated at distant spatial relations, here, entanglement brings with it irreducible relations between the momenta instantiated at distant spatiotemporal locations.

Confined to the low-dimensional image, there is nonseparability. The facts about the quantum state at $x_1 \cup x_2$ are not wholly determined by the states at its subregions. What things are like at x_1 is inextricably correlated with the situation at x_2. But this fact is left out if one just looks at what things are like at x_1 and x_2 individually. To achieve a separable ontology, the wave function realist can postulate a higher-dimensional space in which each point corresponds to an assignment of fields to the total region $x_1 \cup x_2$. The amplitudes are given by the weightings of the different Feynman diagrams. Generalizing from this simple case in which we are just considering two locations, and assuming that the spacetime representation from which we began is continuous, the higher-dimensional space will be continuously infinite-dimensional with each point corresponding to an assignment of field operators to all spacetime points or, assuming discreteness, to the smallest regions in the low-dimensional representation. One thus recovers a separable fundamental metaphysics.

4.4 Interpretations and Interpretational Frameworks

At this stage, we may note that we are no longer considering wave functions on a space with the structure of a classical configuration space as the central elements in the wave function realist's basic ontology. What we have instead is a field defined on another kind of high-dimensional space, one for which locations are correlated with assignments of field operators to regions in a four-dimensional ontology. This is a simple consequence of the fact that the low-dimensional, nonseparable representations from which the separability argument departs are different in the relativistic case than they were in the nonrelativistic one.

What we are seeing is that there is a deeper interpretative framework underlying the wave-function-in-configuration-space view discussed in earlier chapters: that which guides one to a metaphysics for quantum theories lacking fundamental nonseparability and nonlocality. This interpretative framework—call it *Localism* if you like—leads one to adopt a picture of a wave function on a space with the structure of a classical configuration space in the context of nonrelativistic quantum mechanics. But it will lead one to adopt a different metaphysical interpretation for relativistic quantum mechanics and quantum field theories, and for that matter, for nonrelativistic quantum field theories. In this respect, wave function realism provides a framework for interpreting quantum theories that is more flexible than is sometimes recognized.

4.5 Response to Objections

Once we see that the wave-function-in-configuration-space view is only an instance of a broader strategy for interpretation applied to the special case of nonrelativistic quantum mechanics, and that relativistic implementations of this strategy will not rely on configuration space representations, some of the objections to wave function realism canvassed in Section 4.2 may be quickly dispensed with.

A. Recall that the first objection (from Wallace and Timpson) is that wave function realism, when applied to relativistic quantum theories in which particle number is not conserved, leads to an ontologically profligate picture of an infinite sequence of configuration spaces. Viewing the wave-function-in-configuration-space picture as an implementation of an interpretative strategy that applies only to nonrelativistic quantum mechanics, this objection does not get off the ground. The wave function realist need not and should not offer the wave-function-in-configuration-space picture as an interpretation of relativistic quantum theories.

There would be something to this objection were the wave function realist not able to suggest an alternative picture applicable in the case of relativistic quantum theories. But, as we saw in Section 4.3, these concerns are unfounded. In the case of relativistic theories, the wave function realist can start from the same place as the spacetime state realist, with a nonseparable metaphysics of operator assignments to spacetime regions, one that relies on Fock space representations to handle variation in particle number, and reason from there to a more fundamental, separable metaphysics in a high-dimensional space in which each point corresponds to a total assignment of operators to local spacetime points or regions. Wallace argues that:

> wavefunction realism seems to rely on features of toy NRQM which, far from being universal features of any realistic quantum theory, drop away as soon as we generalise. (Wallace 2020)

The response is that wave function realism applied to toy nonrelativistic quantum mechanics relies on features of toy nonrelativistic quantum mechanics. However, the wave function realist achieves an interpretation of more realistic quantum theories by relying on the features of these more realistic theories, those same features on which the spacetime state realist bases his interpretation.

B. The second objection Wallace and Timpson (2010) raise for wave function realism is that the theory "obscures the role of spacetime in quantum theories." What seems to be underlying this objection is a kind of normative constraint that metaphysical interpretations of physical theories should themselves remain close to the mathematical structure of those theories. Wallace and Timpson defend their own spacetime state realism for the fact that "it adds no additional interpretational structure (given that the compositional structure of the system is, *ex hypothesi*, already contained within the formalism); and it gives an appropriately central role to spacetime" (2010, p. 712).

Myrvold presents similar concerns as well noting that characterizing the ontology of quantum field theories in terms other than an assignment of values to spacetime regions "is not what is done in the usual presentations" (2015, p. 3271). He also notes that "the empirical content of [QFTs] remains tied to observables that are associated with regions of spacetime" (2015, p. 3271).

The first thing to note is that the wave function realist does not reject the truth or empirical adequacy of the spacetime representations that are characteristic of usual presentations of both conventional and algebraic quantum field theories. Her claim is not that these representations should be rejected, but rather that they should be seen as metaphysically explained or grounded in terms of a more fundamental representation of a field in a high-dimensional space. This is a metaphysical claim motivated, as we have seen, by a commitment to fundamental separability and the thought that nonlocal correlations should be explained in terms of a more fundamental picture lacking them.[13] Most of the time, the wave function realist takes as a given the truth of spacetime representations, only questioning their fundamentality, so there is no problem with her using them to understand the empirical content of quantum field theories or even their role in standard expositions of those theories.

How closely a fundamental metaphysical interpretation of a physical theory should adhere to the picture immediately suggested by the formalism is a vexed issue. Insofar as this is a desideratum, we can see that wave function realism when applied to nonrelativistic quantum mechanics does better at achieving it than does the strategy when applied to relativistic quantum

---

[13] Although, the kind of spacetime state realism Wallace and Timpson prefer appears to achieve locality as well. The wave function realist appeals to locality rather to distinguish her metaphysics from other interpretational frameworks, such as that of standard Bohmian mechanics.

theories, since textbook discussions of nonrelativistic quantum mechanics often rely on the wave-function-in-configuration-space picture. However, in the relativistic case, the wave function realist does have a justification for moving beyond the sort of picture suggested by the relativistic quantum formalism. What she is trying to do is seek out what the strange observations constituting our evidence for quantum theories are telling us about the fundamental structure of our world. Of course it makes sense, when we construct quantum theories that do justice to our evidence and are relativistically covariant, that we start with spacetime representations. However, persistent correlations that we observe between spatiotemporally distant events suggest that there is something more basic underneath the spacetime picture. The aim of wave function realism is to spell out what this more basic picture looks like if nonlocal correlations do possess a more fundamental explanation.

C. The third critique Wallace and Timpson raise for wave function realism is that relativistically covariant quantum theories violate narratability: the history of events cannot be told as a single story evolving over time, but will vary depending on the way one slices up the spacetime manifold into times. Wallace and Timpson note that this violation is surprising from the point of view of a separable picture like wave function realism, but not from a nonseparable picture like spacetime state realism. To see how narratability fails to obtain for relativistic quantum theories, let's consider one of the simple cases presented by Albert (2015).

Albert describes a system S consisting of four spin-1/2 particles. From the perspective of a frame of reference K, particles 1 and 2 are permanently located at their respective spatial positions, and particles 3 and 4 move with uniform velocity along parallel trajectories intersecting the paths of particles 1 and 2 at spacetime locations P and Q respectively (Figure 4.5).

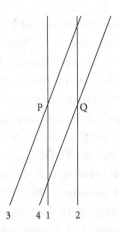

Figure 4.5. Narratability Failure

P and Q are simultaneous from the perspective of K. Albert stipulates that the spin state of the particles is $\psi_{12}\psi_{34}$, where:

$$\psi_{AB} = \frac{1}{\sqrt{2}} \, |\text{x-up}\rangle_A \, |\text{x-down}\rangle_B - \frac{1}{\sqrt{2}} \, |\text{x-down}\rangle_A \, |\text{x-up}\rangle_B$$

Albert notes that from the perspective of K, there is no difference between the following two Hamiltonian descriptions of the joint system's evolution. The effect of the two Hamiltonians may be summarized simply as:

$\hat{H}_1$: S evolves freely

$\hat{H}_2$: Particles exchange spins upon contact.

Interestingly, although reference frame K recognizes no difference between evolution according to $\hat{H}_1$ and $\hat{H}_2$, since from the perspective of K, the state of S is always $\psi_{12}\psi_{34}$, this is not so in any other frame of reference K'.

And so the story or "narrative" of the evolution of S over time is dependent on one's reference frame. Nor can one transform between the description of K and other frames by a Lorentz transformation. No such transformation is possible since it would require mapping identical states in K to distinct states in K'. And so, it looks like specifying a system's state at all times in any one frame of reference (e.g., in K) is not sufficient to specify all facts about that system.

Wallace and Timpson argue that if one adopts spacetime state realism, narratability failure becomes natural. This is due to the interpretation's acceptance of nonseparability:

> Suppose that we have *any* spacetime theory which (i) is nonseparable, so that there can be simultaneous spacetime regions A and B such that the state of A∪B is not determined by the states of A and B separately, and (ii) is also covariant. Covariance entails that there can also be *non*-simultaneous spacetime regions whose joint state is not fixed by their separate states. This opens up the possibility of failure of narratability: specification of global states on elements of one foliation on their own will not in general fix the joint states of non-simultaneous regions. (2010, p. 721)

Let's see in more detail how this is supposed to work. Start (1) by assuming nonseparability so that there are regions A and B that are simultaneous according to some reference frame K and the facts about A∪B are not wholly determined by the intrinsic facts of A and B taken separately. Then, (2) assuming covariance, there is another frame K' according to which A and B are not simultaneous. Suppose (3) that according to K', B is in the future of A. The facts about A and B individually do not determine the facts about A∪B. So in particular, (4) the facts about A do not determine the facts about A∪B. The failure of narratability is a situation in which (5) the fact of whether some event x follows another y, or the chances of y's following x, is not absolute, but depends on one's reference frame. This could be, for example, the facts about A not determining whether B follows or with what chances it is likely to follow from A. So (4) is interpretable as an example of (5). Narratability failure is thus something one might expect given the assumptions of nonseparability and covariance. At least this is what Wallace and Timpson appear to have in mind.

The key issue though, if narratability failure is to provide a case for spacetime state realism over wave function realism, is whether the implication works in the other direction, that is, does the failure of narratability suggest a failure of separability? And does it do so in a way to suggest the failing of wave function realism?

It is not clear why it should. In demonstrating the failure of narratability, one implicitly relies on spacetime representations in which one can make sense of the notion of relativistic covariance. One might argue that the fact that there is no objective distinction between simultaneous and non-simultaneous spacetime regions means one can argue from narratability failure (and covariance) to failure of separability, just as we showed above one can argue from the failure of separability (and covariance) to the failure of narratability. But this would only get us to nonseparability in spacetime.

As we have seen, the wave function realist believes there is a more fundamental metaphysics underlying spacetime representations. And her view is that what manifests itself as nonseparability in ontologically derivative spacetime representations is a more fundamental metaphysics that is separable. So, there is no direct argument from narratability failure to fundamental nonseparability of the kind that would undermine wave function realism. Even if narratability failure implies spacetime nonseparability, spacetime nonseparability is compatible with both fundamental nonseparability (spacetime state realism) and fundamental separability (wave function realism).

The key is to keep the issue of separability/nonseparability in the derivative low-dimensional space straight from the issue of separability/nonseparability in the more fundamental high-dimensional space. As we saw in Chapter 3, the wave function realist thinks there is separability in the latter, but not in the former. Ultimately this means that the explanation of narratability failure Wallace and Timpson find in spacetime nonseparability is available to the wave function realist as well, if she wants it.

D. We may now turn to Myrvold's objection to wave function realism that in quantum field theories, wave functions are constructed from structures defined on spacetime, and so cannot be more fundamental than spacetime structures. Myrvold demonstrates the derivation of wave functions in the context of both nonrelativistic and relativistic quantum field theories to show how the wave function metaphysics is, in the context of actual physical theories, an effective metaphysics merely, not fundamental.

We need not get into the details of these derivations though to see that this criticism of wave function realism is misplaced. First, the fact that wave function representations may be mathematically derived from spacetime representations does not show anything about the direction of ontological priority. The wave function realist (at least in the nonrelativistic context) only argues that wave functions are ontologically prior to spatiotemporal structures. Myrvold's demonstrations do not show that spacetime representations cannot similarly be derived from wave function representations, or the kinds of representations I have argued the wave function realist should take to be tracking fundamental structure in the case of relativistic quantum theories.

But second, as I've tried to argue above, the wave function realist need not deny that the metaphysics of wave functions in configuration space is an effective metaphysics. She may concede that this picture arises only when one subjects what is a more fundamental metaphysics to a nonrelativistic idealization. Indeed, it is one of the main claims of the present chapter that this is precisely the case and I have made suggestions for how the wave function realist may take steps toward articulating more fundamental metaphysics for quantum field theories (Section 4.3). So viewing the wave-function-in-configuration-space picture as an effective metaphysics does not require invoking spacetime as a fundamental background.

It is also worth noting in this context that the sort of higher-dimensional state spaces that play a central role in informing the ontology of the wave function realist are also central in many approaches to what are aspiring to be even more fundamental theories of quantum gravity. It is often argued that in many if not all of the most promising approaches to developing a quantum theory of gravity, spacetime fails to be fundamental (Smolin 2006, Huggett and Wüthrich 2013). Spacetime representations are rather derived from more fundamental nonspatiotemporal representations. Often the more fundamental representations are higher-dimensional state spaces of the kind one finds in quantum theories (for some clear examples, see Baez 1999, Cao, Carroll, and Michalakis 2017). Spacetime then is seen as a derived reality in situations in which these more fundamental structures instantiate the type of symmetries taken to be constitutive of spacetime. This lends support to the idea that some form of quantum state space realism is not an approach to interpretation that is limited in application to toy or idealized quantum theories.

E. The final objection we will consider challenges the wave function realist's privileging of the position representation, and notes that such representations fail to be straightforward in relativistic quantum theories. This is one of the main challenges Wallace raises in his 2020 "Against Wave Function Realism." My response to this objection is similar to the response I have argued the wave function realist should make to the first objection. This is that the objection mistakenly assumes that all features of the interpretation the wave function realist gives for nonrelativistic quantum mechanics will carry over to interpretations of relativistic theories. As illustrated in Section 4.3, just as the wave function realist bases her interpretation of nonrelativistic quantum mechanics on standard presentations of nonrelativistic quantum mechanics, the wave function realist will base her interpretation of quantum field theories on standard presentations of quantum field theories. In the absence of a Lorentz covariant position representation for a quantum field theory, the wave function realist will construct her higher-dimensional representation using a different kind of basis. The example above used a momentum basis. Importantly, the wave function realist may start from the same formal framework as the spacetime state realist.

We may see that at least some of the main objections raised to wave function realism regarding the putative difficulties of extending this ontological framework to relativistic quantum theories rest on the false assumption that the resulting interpretation of relativistic quantum theories must postulate a field on configuration space. Others rest on the failure to see that the wave function realist allows the truth and/or legitimacy of spacetime representations, claiming these are determined by the higher-dimensional facts she posits. So, it seems to me there are no clear barriers to extending wave function realism to the relativistic domain. Although the wave function-in-configuration space view is really only plausible as an interpretation of simple, nonrelativistic quantum mechanics, it is a particular case of a broader framework for interpretation applicable to relativistic quantum theories as well.

4.6 Wave Function Realism in the Limit
of Physical Theorizing

We have now shown how the wave function realist's general strategy of uncovering the more fundamental separable (and local) ontologies that underlie the nonseparable and/or nonlocal low-dimensional ones we encounter in our observations may be generalized to the case of less idealized, relativistic theories. Yet, one might still ask if this strategy for interpretation is one that we should think will be reasonable to apply in the limit of inquiry, in a context in which we are not looking for an interpretation of what is merely a less idealized quantum theory, but the final quantum theory.

I would argue that we do not have positive reasons for believing there ever will be such a thing as a final quantum theory. We have quantum theories that work very well, astoundingly well for many purposes, but there is no good inductive argument that takes us from this fact to the conclusion that there will ever in fact be a completed or final (and true) quantum theory, or any completed physical theory for that matter (see Ney 2019). Indeed, given the large open problems still remaining (Smolin 2007), there is reason to think that physics may have to undergo several revolutions before it may reach anything like completion, revolutions that may very well be so radical as to make the resulting scientific theory something we would not recognize as a physics, let alone a quantum physics. Even if there will someday be a completed physics, we can't be very sure what this will look like, and so the best we can do now is interpret the physical theories we have using methodological principles that have worked well for us in the past to support scientific understanding. The best physical theories we have at this time are quantum theories, and of these, I have argued the best permit a wave function realist interpretation, one that may be favored for its ability to produce pictures of the world that are separable and local, and so comprehensible in a way other interpretations are not.

However, there may be a reason to think that wave function realism is a framework for interpreting quantum theories that will not be able to reveal what our world is ultimately like for a reason that has not yet explicitly come up here, namely, that wave function realism yields interpretations of quantum theories that fail to be basis- or coordinate-free.[14] In using configuration space representations to inform their choice of a fundamental ontology for nonrelativistic quantum mechanics, wave function realists privilege a position basis. Their choice of other frameworks for quantum field theories will inevitably privilege other kinds of bases, momentum or Fock state bases. But, bases, like coordinatizations, are not generally viewed as part of objective structure. Rather, bases are tools imposed on whatever objective structure there is to facilitate a kind of representation that is convenient for some purpose. And so, the argument goes, we should not regard the ontologies favored by the wave function realist as fundamental. The ontologies favored by the wave function realist should be regarded as less than wholly objective. Instead, we should prefer ontological frameworks that represent the fundamental quantum state in a basis-independent way.

We only very briefly discussed state vector representations in Chapter 1. However, those who press this concern for wave function realism want to suggest that a better view of quantum states, for the purpose of informing fundamental ontological structure in a basis-independent way, is a representation in terms of state vectors, or density operators defined in terms of these state vectors. Rather than viewing the world fundamentally as field spread out over a high-dimensional space, with each point corresponding to a configuration of particles, or arrangement of fields, the idea is to represent the world by a state vector in Hilbert space, and have this representation guide one's fundamental ontology. Hilbert space is just a linear vector space on which a notion of inner product of vectors is defined. There are at least two different ways of implementing this.

---

[14] Thanks to David Wallace for pressing me on this point.

The most straightforward is what is advocated by Sean Carroll and Ashmeet Singh in their article "Mad-Dog Everettianism":

> What is fundamental does not directly involve spacetime or propagating quantum fields, but simply a vector moving smoothly through a very large-dimensional Hilbert space. (Carroll and Singh 2019 p. 103, see also Carroll 2019)

This is what we may call *the ray-in-Hilbert-space-view*. It may be seen as an alternative version of wave function realism, a version that interprets the wave function not fundamentally as a field smeared out over a high-dimensional space, but instead as a ray fixed onto a particular point in Hilbert space, a different kind of high-dimensional space. The coordinates of the world's state vector may itself be described in terms of different bases, for example the position or momentum basis, but the vector has a location in its Hilbert space that is independent of such bases. Note how different this ontology is from the one we have been assuming up until now. As we saw in Section 1.1, position and momentum wave functions, as the wave function realist understands them, have distinct objective features. While the position wave function for a quantum system may be narrow and peaked, the momentum wave function for that same system may be flat and spread out. Wave functions, if they are instead understood as state vectors, are basis-independent. They take the same shape independent of the bases we impose on them.

An alternative way of implementing the "basis-independent" strategy is to take there fundamentally to be a single density operator yielded by the correct universal state vector, and assign this density operator to spacetime as a whole. This may then be interpreted as the fundamental world-property assigned to spacetime. Or, if one, like Wallace and Timpson, thought this yielded too lean an ontology (see Section 2.6), one could accept as additional bits of fundamental ontology all subregions of spacetime, and take these subregions to bear additional fundamental properties given by the appropriate reduced density matrices. This would yield spacetime state realism. Note that spacetime state realism, like the ray-in-Hilbert-space view is also "coordinate-free" in the relevant sense, as these reduced density matrices are arrived at using the basis-independent partial trace operation.

So why should the wave function realist prefer her framework for interpreting quantum theories, rather than one of these "coordinate-free" alternatives? The wave function realist does not, as we've seen, reject the claim that state vectors and density matrices provide approximately correct representations of quantum states. The question is which kind of representations do the best job of revealing the fundamental ontologies we should assign to quantum theories. In general, when interpreting quantum theories, we are met with four kinds of representation:

1. Macroscopic representations in space or spacetime
2. Microscopic representations in space or spacetime
3. Wave function representations in some higher-dimensional space, e.g. for non-relativistic quantum mechanics, representation of a wave function in a space with the structure of a classical configuration space
4. State vector representations in Hilbert space

The claim of the wave function realist is that the representations at levels (1)–(3) may each provide ontologies that at least approximately track what there is, with (3) being the best level to represent what our world is fundamentally like according to quantum theories. The question is why wave function realists shouldn't think that level (4) representations provide even better representations of fundamental ontology, because they are basis-independent representations.

The wave function realist rejects this claim for two reasons. The first is that, as we have already seen, the argument for wave function realism is that we ought to privilege interpretations of quantum theories that are separable and local. And both spacetime state realism and the ray-in-Hilbert-space view fail to have metaphysics that are separable. According to spacetime state realism, there are facts about properties of spacetime regions that are not determined by the properties of their subregions. The ray-in-Hilbert-space view fails to be separable because it lacks in the first place an ontology of distinct objects occupying nonoverlapping regions. Actually, as an interpretation of wave function realism, this makes the ray-in-Hilbert-space view seem unmotivated. We saw in Chapter 2 that there are many alternative approaches to interpreting the wave function that can capture the facts of quantum entanglement. If wave function realism cannot claim to provide one with a fundamentally separable and local metaphysics, then what reason is there to prefer this view over the others?

Second, the shared project in which Carroll, Timpson, Wallace, and I are engaged is the project of looking for an account of what quantum theories may be suggesting about the world around us. This is a world that at least seems to include material objects, living organisms like you and me, physical artifacts like tables and chairs, celestial bodies like stars and planets, and so on, objects that seem to occupy definite spatiotemporal locations, having shapes and sizes and life spans. And so somehow, our fundamental picture of the world according to quantum theories, if it is to be a fundamental picture of the world around us, must provide the building blocks for a world with macroscopic material objects. If we are going to take seriously a fundamental ontology for quantum theories, we must find some way of demonstrating how that ontology may ultimately constitute the macroscopic objects that we already know exist.[15] Although the wave function realist doesn't question that we can, and physicists do, provide mappings between Hilbert space representations and macroscopic representations, the project of

---

[15] Or provide a compelling error theory that can explain how and why we falsely believe these objects (including ourselves) exist.

establishing an ontology as fundamental requires more than just such mappings. What more is required will be discussed in detail in Chapter 6, however for now I will just note that the wave function realist is skeptical that defenders of these alternative Hilbert space approaches can carry this project out. This is most clear for the ray-in-Hilbert-space view where it is commonly viewed as puzzling how a ray in Hilbert space could have sufficient structure to build up or constitute material objects.[16]

However, the concern is not wholly about there being sufficient structure in such representations. The question is what does the structure there is in Hilbert space representations have to do with the world we are trying to describe. This point strikes equally against spacetime state realism as does it against the ray-in-Hilbert-space view. Wave function realist interpretations privilege certain bases, but they also (as I will argue in Chapter 7) have a way of showing how a fundamental quantum ontology can come to constitute a world of tables and chairs, people and pointers. We might prefer in principle to be more egalitarian, but the trouble is that these more egalitarian frameworks are so abstract as to lose the constitutive connections to the world we wanted to understand. So, I argue, the fact that wave function realism isn't basis-independent is not a strike against it. The wave function realist's claim that the world is fundamentally a field spread out in high dimensions over points corresponding to ways some objects or a field may be spread out in lower dimensions is crucial to its ability to recover the empirical image of our world.

[16] Wallace and Timpson (2010), p. 701 and North (2013), p. 191.

5

Must an Ontology for Quantum Theories Contain Local Beables?

5.1 The Constitution Objection

We have now seen the central case for wave function realism. I haven't tried to convince the reader that wave function realist interpretations of quantum theories are definitively the answers to questions about the deep structure of reality. For we don't have the final quantum theory at hand, only a collection of relatively fundamental quantum theories that work well for many purposes. The best we can do at this time is to offer interpretations of what these theories suggest about the world. Furthermore, the theoretical and pragmatic considerations that favor wave function realism over other ontological frameworks fail to conclusively settle the issue of which is the right approach to interpretation. And so we must accept a certain degree of modesty in pursuing one approach to interpretation over the others. I regard wave function realism as an approach that may provide relatively straightforward interpretations of the quantum theories we find in today's physics, interpretations that have the virtue of being to a great degree simple and comprehensible, at least as a matter of the kind of fundamental ontology they describe. For these reasons, I regard wave function realism as worth pursuing.

However, there exists a class of objections to wave function realism stating that even if we allow that there may be something straightforward and comprehensible about the fundamental ontologies for quantum theories proposed by the wave function

The World in the Wave Function. Alyssa Ney, Oxford University Press (2021). © Oxford University Press.
DOI: 10.1093/oso/9780190097714.003.0005

realist (to the extent that such ontologies are separable and fail to contain any unmediated action at a distance), there is a problem with these ontologies in that they cannot serve as the constitutive foundation for the world we see around us. And this undermines the promise of wave function realism to serve as a framework for the interpretation of quantum theories. For even if a world fundamentally consisting of a wave function evolving in a high-dimensional space is comprehensible, such a world cannot be *our* world. After all, *we* and the objects we interact with could not be made out of wave function stuff.

The rough form these criticisms take may be viewed as a three-premise argument:

Premise 1: An ontology for quantum theories must be capable of constituting macroscopic objects, including ourselves and the objects we perceive around us.

Premise 2: To make it plausible that it may do so, an ontology for quantum theories should include a class of fundamental local beables, that is, entities assigned to definite locations in three-dimensional space or spacetime.

Premise 3: The ontologies for quantum theories recommended by the wave function realist do not include fundamental local beables.

Therefore,

Conclusion: Wave function realism is misguided as a framework for interpreting quantum theories.

I will refer to this argument as *The Constitution Objection*.

The goal of this chapter and those that follow will to be to grapple with the main versions of this objection that have been put forward in the quantum ontology literature. As we will see, I think there are some plausible ways of providing justification for the first premise. Indeed, the main challenge that I will address on behalf of the wave function realist in later chapters is the challenge of providing an account of the existence of the macroscopic objects we interact with, given the assumption that the world is fundamentally constituted by a wave function in a high-dimensional space. I think this can be done. It is really the second premise, which says that providing a plausible account of the constitution of macroscopic objects demands the postulation of fundamental local beables, of which I am skeptical. Nonetheless, I do think that there are problems with the various ways the constitution objection has been developed in the literature even before we get to the second premise. I will articulate these concerns shortly. First, however, I want to say a word about my framing of the third premise, which I have intentionally stated in such a way that it may be seen as a direct (analytic) consequence of what it is to be a wave function realist.

Wave function realism is a general framework for the interpretation of quantum theories that takes these theories to be describing a world in which there is fundamentally a field (the wave function) on a high-dimensional space, and perhaps additionally some other objects described by other variables in the quantum formalism. These too are assumed to inhabit the high-dimensional space of the wave function. The wave function realist thus rejects the claim that the fundamental ontology of quantum theories consists of low-dimensional objects scattered in three-dimensional space or spacetime, or some properties of these objects. This is what I intend when I say that the ontology for quantum theories recommended by the wave function realist does not include any fundamental local beables.

Now an important issue when it comes to filling out the nature of the world according to the wave function realist concerns whether these fundamental objects of quantum theories may come to constitute anything else, in particular, whether a wave function in a high-dimensional space may come to constitute, as derivative existents, anything like the low-dimensional objects of our ordinary experience. As has been mentioned in previous chapters, the wave function realist usually thinks that it can. However, the wave function realist will not regard such low-dimensional objects as part of the *fundamental* ontology for quantum theories. Typically, the wave function realist will agree with those who raise the constitution objection that the world (our world, a quantum world) is one in which there are low-dimensional macroscopic objects including ourselves and the objects with which we interact. The question is, what in the quantum world are the more fundamental entities that may be said to constitute these objects? With the third premise thus clarified, we may now consider what may be said in defense of the first two premises of the constitution objection.

5.2 Doing without Macroscopic Objects

The first premise of the constitution objection, that an ontology for quantum theories must be capable of constituting macroscopic objects, including ourselves and the objects we perceive around us, has been defended in several ways by philosophers in critiques of wave function realism. But we may note that this is not a premise all interpreters of quantum theories must accept.

First, one might argue that the lesson one should draw from the developments in physics that led to quantum theories is that the material world that exists around us is much stranger than we previously thought. It is so strange that one ought to reject one's own existence, as well as the existence of all other macroscopic objects. One should believe that only the world directly described by quantum theories, the wave function in its high-dimensional space, is real. All else is an illusion.

Or, one might reject the first premise by upholding a kind of dualism, placing the wave function and other high-dimensional physical objects on one side of a metaphysical divide, and ourselves and the world of macroscopic, material objects on the other. In this way, one might pursue the question of the correct ontology for quantum theories without having to worry about how to see the wave function as constituting the macroscopic world of our ordinary experience. However, it is hard to see what might justify this view of the world, that both of these ontologies exist and yet there is no constitutive relation between them.

An approach more grounded in traditional philosophical doctrine takes a path midway between the two previous strategies for rejecting the claim that a quantum ontology must be capable of constituting macroscopic objects. This would be to allow that ourselves, our minds and experiences, exist, as do the entities described by quantum theories, but at the same time insisting that macroscopic material objects do not. This would be to adopt a position looking in some ways like a traditional mind-body dualism. One could then argue that minds and mental phenomena are different kinds of things than what is described by physics, perhaps constituted out of some distinct kind of spiritual substance, philosophical ectoplasm. One would thus avoid the need to see ourselves or anything else as constituted out of the stuff of quantum theories. This view would be unlike the traditional mind-body dualism of Descartes in denying the existence of macroscopic material objects. In this way, the position is still somewhat radical and out of step with philosophical tradition. But it would not be so radical as to also deny our experiences of macroscopic material objects, which would be a consequence of the first way of rejecting the first premise.

I should note, one doesn't have to take on such radical views in order to reject the first premise of the constitution objection. Indeed, there is a way of doing so that is more in line with recent trends in the philosophy of science. This involves regarding the claim that a quantum ontology must be such as to constitute ordinary macroscopic objects as relying on a dubious assumption about the unity of science. This is the assumption that our best physical theories ultimately serve to give approximately correct and fundamental accounts of the kinds of phenomena that exist in all domains, that physics is a fundamental science.

Note that this assumption underlying the first premise is strictly speaking weaker than what is often called (and dismissed as) reductionism. For the first premise of the constitution objection rests only on the assumption that physics is more fundamental than the other sciences in the sense that somehow or other its ontology may serve as a constitutive basis for the entities of all of the other sciences, not on any claim about the reduction of one ontology to another. Reductionism is often taken to require something quite strong, that the truths of the other sciences can all be explained from within physics, or sometimes, that the entities and kinds of all other sciences must somehow be identified with those of physics or discarded.

As I said, even though the assumption about the unity of science underlying the first premise is weaker than reductionism, this idea that the entities of physics are special in forming a basic class of entities out of which all others are constituted has been questioned. One influential critique comes from Nancy Cartwright (1999) who labels as "fundamentalist" any hierarchical picture according to which physics gives us a set of principles that can correctly characterize what all things are fundamentally like in all circumstances. Cartwright's argument stems from considerations about the very contrived circumstances in which the claims of our fundamental physical theories have been tested and confirmed, noting that these are very much unlike the kinds of situations in which we ordinarily observe the regular behavior of macroscopic objects.[1] In those latter kinds of situations, ones in which we best model systems using the principles of Newtonian physics or folk psychology,

---

[1] She says: "Perhaps we feel that there could be no real difference between the one kind of circumstance and the other, and hence no principled reason for stopping our inductions at the walls of our laboratories. But there is a difference: some circumstances resemble the models we have; others do not. And it is just the point of scientific activity to build models that get in, under the cover of the laws in question, all and only those circumstances that the laws govern. Fundamentalists want more. They want laws; they want true laws; but most of all, they want their favourite laws to be in force everywhere. I urge us to resist fundamentalism. Reality may well be just a patchwork of laws" (1999, p. 34).

there are no grounds for thinking that the laws of quantum theories apply. There are no grounds for thinking these medium-sized bodies or minds are at some more fundamental level of description the stuff of quantum theories.

Both wave function realists and their critics have tended to agree for the sake of discussion that we may see physics as describing what is the constitutive basis of the ordinary, macroscopic objects we encounter in the world, including ourselves. There is an important debate that continues as to whether this is right, or whether as Cartwright and others argue, we should regard physics as constituting merely another "special" science, as opposed to a general science that succeeds at describing everything whatsoever including ourselves. But although to accept Cartwright's critique of fundamentalism would provide the wave function realist an easy way to evade the constitution objection, it is not a way of doing so that I will pursue here.

Indeed, I will not suggest we reject the claim that an ontology for quantum theories must be capable of constituting macroscopic objects, including ourselves and the objects we perceive around us. What I will challenge in this book is instead the assumption made by critics of wave function realism that to make it plausible that a quantum ontology can provide a constitutive basis for the existence of macroscopic objects and ourselves, it must include a fundamental ontology of local beables.

This is not to say I believe that critics of wave function realism have accurately presented the reasons why we should think that an ontology for quantum theories must be such as to be capable of constituting macroscopic objects. There are two main arguments critics of wave function realism typically provide for this claim, the first premise of the constitution objection. The next two sections in this chapter (5.3 and 5.4) capture my reasons for rejecting these arguments. Section 5.5 articulates what I take to be a safer way of defending the claim that an ontology for quantum theories should be capable of forming a constitutive basis for macroscopic objects.

Chapters 6 and 7 then develop what I regard as the best way of rejecting the constitution objection's second premise, by showing how macroscopic objects may be constituted out of a quantum ontology that lacks fundamental local beables.

I will say, before presenting these arguments for the first premise of the constitution objection, that these arguments are frequently conflated. This happens, I think, because they have been defended by a group of opponents to wave function realism (Tim Maudlin, Detlef Dürr, Shelly Goldstein, Nino Zanghì, Valia Allori, and Roderich Tumulka) who agree about quite a lot. Indeed it may be, to the extent that I am capturing their arguments correctly, that each would agree with both of the two arguments I am about to outline. Still, I hope it can be seen clearly that there really are two distinct arguments that may be teased out of these authors' work. And so these arguments require different responses on behalf of the wave function realist.

5.3 The Threat of Empirical Incoherence

The first, very intriguing way of defending the first premise of what I call the constitution objection to wave function realism starts with a general view about the way confirmation of theories by evidence works. This view seems, on the face of it, very simple. A theory makes predictions that certain objects will be observed to have certain features. And then we observe those objects to have those features.

Maudlin argues that this framework for empirical confirmation raises more of a challenge to some proposed physical ontologies than to others:

> In order to be of interest, physical theories have to make contact with some sort of evidence, some grounds for taking them seriously or dismissing them. . . . [In classical physics], the contact

between theory and evidence is made exactly at the point of some local beables: *beables that are predictable according to the theory and intuitively observable as well.* (2007a, pp. 3159)

I take Maudlin to be suggesting here that for theory confirmation to succeed in the way it does in the case of classical physics, there should be an overlap between the kinds of objects the theory says exist, those things it makes predictions about, and those we observe that constitute the confirming evidence. Call this *the overlap principle.* We can use the overlap principle then to argue that if the evidence for our quantum theories consists of macroscopic objects having certain features, then for it to be clear that our quantum theories are justified, they should make predictions about, that is, have an ontology that constitutively includes, macroscopic objects with those features. And if, as Maudlin thinks, our evidence for quantum theories consists of local beables, then for it to be clear that our theories are justified, they should make predictions about, that is, have an ontology that includes, local beables.

Before continuing, we should note that the overlap principle is not a principle that is plausible for all scientific theories whatsoever. Its use here rests on a special assumption about quantum theories, namely, that they aspire to a kind of ontological completeness, at least insofar as they aim to capture constitutive facts about the objects around us. In most cases of scientific confirmation, it will be sufficient for a theory to be confirmed that there be some stable correlation or connection between the ontology of the theory, what it makes predictions about, and the evidence for it. One need not require literal overlap between the ontology of the theory and the evidence for it. For example, meteorology makes predictions about changes in temperature, but it may make no predictions about the movement of mercury up a glass cylinder. Nonetheless, we use glass cylinders containing mercury in order to confirm meteorological predictions. In this case, there isn't an overlap between the ontology of the theory (meteorology) and the evidence for it. There is only a correlation. But this failure of overlap is no problem because we have another part of science, a more complete theory, that is able to connect the predictions of the theory, increases in temperature, with our evidence, the rise of the mercury, to explain why there exists such a correlation. This more complete theory has an ontology that includes both the ontology of and the evidence for meteorology. We can thus see that what the overlap principle really applies to are those theories that aspire to provide more complete accounts of our world in the sense that constitutive facts about all of the objects around us will be captured by the facts of that theory. In the case of such theories, if there fails to be overlap between the evidence for the theory and its ontology, in particular, if the evidence appears to outstrip the ontology, then there is a problem: if the theory (under that ontological interpretation) were true, it seems the evidence for it would fail to exist.

The hope is that at least some quantum theories (of today? of the future?) have the capacity to provide accounts of the constitution of the objects around us. The overlap principle thus appears to license a claim about the ontological interpretation of these theories. For it to be clear these theories are justified, one must interpret them so that the evidence for them is contained within their ontologies. Otherwise, to use Jeffrey Barrett's (2001) phrase, such interpretations will be *empirically incoherent*: they will be such that were they to be correct, the theories they are interpretations of would fail to be justified. Most agree with Maudlin that the evidence for quantum theories consists of local beables, more specifically, macroscopic objects with determinate spatiotemporal features—we observe pointers pointing to definite locations, tracks in cloud or bubble chambers or on computer screens, clicks of Geiger counters at determinate times, and the like. For this reason, an ontological interpretation of a quantum theory that fails to include macroscopic objects with determinate spatiotemporal features appears to face a charge of empirical incoherence.

Maudlin argues that empirical coherence is more of a challenge for some frameworks for interpretation than others. When it comes to frameworks like wave function realism that reject fundamental local beables, he worries that this "is going to be a much, much more complicated business than understanding a theory with local beables" (2007a, p. 3160). But is the contact between theory and evidence really so much more complicated for ontological interpretations of quantum theories that lack local beables?

Generally, wave function realists have tended to accept Maudlin's claim that our evidence consists of macroscopic local beables. They reply to his argument that on their view, these beables are part of the ontology of quantum theories. They are simply derivative entities constituted out of wave function stuff. This allows these wave function realists to say that there is still a constitutive overlap between theory and evidence since these macroscopic local beables are constituted by the wave function. However, this raises for them a challenge of how we may see local beables as constituted out of the wave function. This is a vexed topic, one we will take on in the following two chapters. What I want to argue here, however, is that there is a more straightforward and less contentious way of responding to Maudlin's argument, one that does not require the wave function realist to demonstrate that local beables may be constituted out of the wave function.

One believing an interpretation of quantum theories according to which they tell us that the world around us is fundamentally a world of local beables will take our evidence for quantum theories to consist of local beables. In this way, the contact between theory and evidence is straightforward. But one believing an interpretation of quantum theories according to which they tell us that the world around us is fundamentally the high-dimensional world of the wave function will take our evidence to have a different nature.

She will take it to consist of wave function stuff. In this case, the contact between theory and evidence may similarly be straight-forward. My point is, there are many ways of satisfying the overlap principle, as many ways as there are of providing an ontology for a physical theory. The difficulty and complications Maudlin speaks of arise only for those who would claim there fails to be a straight-forward overlap between the ontology of a quantum theory and our evidence for it. The wave function realist need not concede this point.

I allow that, given the way our evidence presents itself, it does seem to involve pointers pointing at one place, here, rather than some other place, there. And for there to be a distinction between here and there, these places to which our pointers point and those to which they do not, there must exist genuine (low-dimensional) spatial locations. But for us to possess the evidence we do, there needn't be things with genuine low-dimensional locations. There must be a distinction (and according to the overlap principle, one captured in the ontology of the theory) between the evidence that confirms a theory and the evidence against it, and the evidence for the theory must exist. So, using our pointer example again, there must be a difference that can be grounded in the ontology of the quantum theory between what we refer to as "the pointer pointing to up" and "the pointer pointing to down." But the wave function realist is free to trace this difference to differences in the state of the wave function in the two scenarios. Empirical confirmation per se does not need to proceed through our observations of objects at distinct three-dimensional spatial locations rather than (what is more general) our observations of different sorts of states in what-ever is the correct ontology.

It may be helpful to contrast the situation for the wave func-tion realist with another case in which there is a legitimate threat of empirical incoherence. This threat arises specifically for canon-ical approaches to quantum gravity and has been discussed by

Richard Healey (2002). Canonical approaches to quantum gravity are those that start from a Hamiltonian formulation of general relativity that then gets quantized. A frequently noted result is that on the standard way of doing this, the time variable drops out of the theory's central dynamical equation. This has suggested to some physicists (e.g., Julian Barbour and Carlo Rovelli) that according to such theories, there is no change and so time is unreal. Healey calls this position, this metaphysical interpretation of particular formulations of quantum gravity, "timeless Parmenideanism." Healey has argued persuasively that since a theory of quantum gravity is supposed to be a fundamental physical theory, such an interpretation of quantum gravity is empirically incoherent. Why? Because what it takes to confirm a theory is for the theory to make a prediction at one time and then see it confirmed at a later time. That is how empirical confirmation works; it is a diachronic process. But if timeless Parmenideanism is correct, then time is unreal and so there is never any confirmation of the theory of which it is an interpretation.

This case raises a genuine conceptual issue about the nature of empirical confirmation. The claim that predictions must occur before the confirming observations has intuitive support: the way we tend to view confirmation involves the development of a theory at one time, followed by the formulation of predictions, and finally (if one is lucky), confirmation at a later time. How could a theory be confirmed before it is formulated? This doesn't seem to make sense. So there is an assumption about temporal order built into our intuitive notion of confirmation.

Moreover, this feature is explicitly built into many theories of empirical confirmation today.[2] Predictivism is the dominant position,

---

[2] Popper's theory of corroboration of theories by data also requires a temporal ordering. Those theories that are corroborated are those that have survived attempts to refute them (1934/2002).

with those working within both the hypothetico-deductive and Bayesian traditions arguing that a theory must first predict a set of data before that data may be used to confirm the theory. Indeed, given the standard Bayesian assumption that in confirmation one conditionalizes on one's total evidence, if the data were already known, then it cannot serve to boost rational credence in a theory. Even those who resist predictivism (for example, those Bayesians who prefer to weaken the total evidence requirement), tend to argue that even if a set of data was not itself originally predicted from the theory (it is a case of "old evidence," like the precession of Mercury's perihelion was for general relativity), for it to confirm the theory it *must have been possible* for it to have been predicted from the theory before the theory was developed. Even this weaker assumption relies on facts of temporal ordering.

Thus, timeless Parmenidean interpretations of canonical approaches to quantum gravity raise a real issue about empirical coherence. If we are going to see them as promising ways of interpreting quantum gravity, they challenge some widely held views about confirmation. This isn't to say that these challenges cannot be met. There is some merit in the idea that the temporal relation between the formulation of the predictions of a theory and the discovery of data shouldn't matter for the theory's confirmation. Instead, what ought to matter is what information the theorist actually uses in constructing the theory. But there is at least an issue here to be discussed based on the assumptions built into our intuitive notion and philosophical theories of confirmation. The same situation does not apply in the case we are considering against wave function realism.

No theory of confirmation involves the requirement that the data for a theory be located in three-dimensional space or spacetime. One might perhaps think that the intersubjective availability of evidence, a requirement for the confirmation of scientific theories stressed by the logical empiricists, requires that our evidence be located in three-dimensional space or spacetime. But if one's evidence is such that other theorists can also acquire it and see that it confirms one's theory, then it is not clear why it should matter in which type of space this evidence is located. One way to ensure that evidence for a theory is intersubjectively accessible is for it to have a location in a common three-dimensional space. But another way is for it to have a location in some other common spatial framework, for example, the wave function's space. The wave function realist is not claiming that our evidence is confined to our own private mental states. And so there is no obvious concern about intersubjective availability, even if a confirmation theorist were motivated to impose such a requirement on the evidence that may confirm a scientific theory.

Now there is still a lingering issue for the wave function realist. Even if our evidence doesn't consist of local beables, but only out of some features of the wave function, to be confident there is such evidence for quantum theories, the wave function realist must be confident that the evidence that we have, the evidence we have described in ways like "the pointer is pointing up" may be constituted out of something like the wave function. This may seem to present a challenge to the wave function realist to spell out the nature of the relationship between what appear to be events like pointer readings and states (or sequences of states) of the wave function.

The answer is that yes, a connection has to be spelled out, but the project of spelling out this connection is going to be much simpler than the project of spelling out how the wave function could metaphysically constitute a three-dimensional pointer. Here all that is required is that we show how there may be a correlation between certain states of the wave function and situations in which it has seemed to us that we are looking at pointers. But as we will see in the next chapter, the way that wave functions were originally introduced into quantum mechanics ensures that such correlations obtain.

5.4 Primitive Ontologies and Local Beables

We may now move on to consider a second justification for the first premise of the constitution objection.[3] This derives not from a claim about the nature of *our evidence for* quantum theories but instead from a claim about *the explanatory targets of* quantum theories. There is a framework for the ontological interpretation of physical theories, including quantum theories, the primitive ontology framework, that, as we've seen, holds that there are certain facts we begin with in physical theorizing, certain assumptions about the kinds of objects physical theories are aimed at describing. In their 1992 paper introducing this framework, Dürr, Goldstein, and Zanghì noted that physical theories, including quantum theories, come with primitive ontologies, classes of entities that these theories are aimed at describing. Moreover, these primitive ontologies include "the basic kinds of entities that are to be the building blocks of everything else" (p. 7). "The elements of the primitive ontology are the stuff that things are made of" (Allori, Goldstein, Tumulka, and Zanghì 2008, p. 362). Thus, according to the primitive ontology framework for the interpretation of physical theories, the explanatory targets of quantum theories are (perhaps inter alia) the entities that constitute the macroscopic entities we see around us. And so, it follows that an ontology for quantum theories must be capable of constituting macroscopic objects.

It should be noted that if what was just described exhausted the notion of a primitive ontology, there would be very little space between this argument for the first premise of the constitution objection and the more simple-minded one I will present shortly, and so I would have very little to disagree with. However, to claim that physical theories bring with them primitive ontologies is to claim more than merely that physical theories have classes of entities that

[3] Many of the points I make in this section descend from joint work with Kathryn Phillips and previously published in Ney and Phillips (2013).

serve as their explanatory targets and that also serve to make up everything else, including macroscopic objects. To make this clear, and see all that comes with the deployment of the notion of a primitive ontology in the debate over the interpretation of quantum theories, let us see in more detail how this notion gets characterized by advocates of the primitive ontology approach:

> According to (pre-quantum-mechanical) scientific precedent, when new mathematically abstract theoretical entities are introduced into a theory, the physical significance of these entities, their meaning insofar as physics is concerned, arises from their dynamical role, from the role they play in (governing) the evolution of the more primitive—*more familiar* and *less abstract*—entities or dynamical variables. For example, in classical electrodynamics the *meaning* of the electromagnetic field derives solely from the Lorentz force equation, i.e., from the field's role in governing the evolution of the positions of charged particles. . . . That this should be so is rather obvious: Why would these abstractions be introduced in the first place, if not for their relevance to the behavior of something else, which somehow already has *physical significance*? (Dürr et al. 1992, p. 7).

> What we regard as the obvious choice of primitive ontology— the *basic kinds of entities that are to be building blocks of everything else*—should by now be clear: Particles, *described by their positions in space, changing with time*—some of which, owing to the dynamical laws governing their evolution, perhaps combine to *form the familiar macroscopic objects of daily experience*. (Dürr et al. 1992, p. 9, my emphasis)

> The PO [primitive ontology] of a theory—and its behavior—is *what the theory is fundamentally about*. It is closely connected with what Bell called the "*local beables*" . . . The elements of the primitive ontology are *the stuff that things are made of*. (Allori et al. 2008, p. 362, my emphasis)

To introduce a PO for a theory means to be explicit about what *space-time entities the theory is fundamentally about*. There are various possibilities for what type of mathematical objects could represent the elements of the PO, including particle world lines as in classical or Bohmian mechanics, world sheets as maybe suggested by string theory, world points as in the GRW theory with the flash ontology. (Allori et al. 2011, p. 4, my emphasis)

We can thus see that there are at least four elements that come with the notion of a primitive ontology, four constraints that a primitive ontology in the sense of Allori, Dürr, Goldstein, Tumulka, and Zanghì must satisfy. They are:

Familiarity: Elements of the primitive ontology have some antecedent familiarity/physical significance prior to the development of the theory.

Relevance (with respect to the theory): The behavior and features of the elements of the primitive ontology are the primary explanatory targets of the theory. The elements of the primitive ontology are those entities the theory is fundamentally about.

Constitution: The elements of the primitive ontology are the things that make up other things in the world of our experience, including ordinary macroscopic objects.

Spacetime location: The elements in the primitive ontology have locations in ordinary space or spacetime. They are local beables in Bell's sense.

Although they never make this list explicit, it is easy to find mention of each of these constraints whenever the notion of a primitive ontology is explicated. And so I propose that we regard 'primitive ontology' as a technical term whose meaning is settled by these constraints.

I should emphasize in case it is not clear already that the notion of a primitive ontology is not the same as that of a fundamental ontology. There is a live dispute today about how we should understand 'fundamental,' but it is certainly something different than what these authors mean by 'primitive.' To see that the concepts of a fundamental ontology and a primitive ontology are not the same, we may note that advocates of the primitive ontology approach do not deny that the wave function should be an element of the fundamental ontology of a quantum theory. They merely argue that the wave function is not primitive and so cannot exhaust the fundamental ontology:

> Thus it seems that for a fundamental physical theory to be satisfactory, it must involve, and fundamentally be about, "local beables", and not just a beable such as the wave function, which is nonlocal. (Allori et al. 2011, p. 10)

The idea is that to capture all of the facts in the domain of quantum theories, it is necessary to include both facts about the primitive ontology and facts about the wave function, just as in classical electromagnetism it is necessary to include both facts about the primitive ontology and facts about the fields. So it is not necessarily the case that all of the fundamental ontology of a physical theory be primitive ontology, even if all of the primitive ontology is fundamental ontology. To be skeptical (as I am) about the claim that quantum theories have primitive ontologies is not thereby to be skeptical that quantum theories have fundamental ontologies.

Let's now try to clarify the argument for the claim that quantum theories should have primitive ontologies. A natural thought given some of what is said in the above passages (especially those from Dürr et al. 1992) is that the motivation for a primitive ontology comes from the fact that there is something already familiar to us, some domain of local beables that constitute the macroscopic objects of our experience, that any adequate formulation of quantum mechanics aims to be about, and this is not the wave function.

This point is in some ways compelling. Of course physicists did not develop quantum mechanics out of a desire to better understand the wave function. Rather wave function representations were used to explain the behavior of something else, something with which we were already familiar, subatomic matter and radiation. But even if one concedes that in a real sense, any adequate quantum theory should be about subatomic matter and radiation, this does not show that any adequate interpretation of a quantum theory must involve a fundamental ontology of antecedently familiar objects that are local beables, a primitive ontology.

There is one sense of "what quantum theories are about" for which certainly this is determined by the historical conditions in which these theories were developed: in this sense, quantum theories are about whatever their founders were interested in explaining. Call this *the explanatory sense* of what quantum theories are about. But there is another way one might interpret the phrase "what quantum theories are about," and that is in terms of what the central representational tools in the formulation of quantum theories denote. Call this *the ontological sense* of what quantum theories are about because, as I will argue, this is the only sense that has direct implications for the fundamental ontology of the theory.[4] What I want to point out is that what quantum theories are about in the explanatory sense (what their founders were most interested in explaining) need not overlap with what they are about in the ontological sense. And so even if it is true that a physical theory's ultimate explanatory target must be something with antecedent familiarity and significance to us, this thing need not occur as part of the fundamental ontology of the theory.

Consider, for example, a theory of the chemical properties of molecules. If this theory is successful, one must get out of it an explanation of why molecules of the kind we observe exist and

---

[4] This is not to say that the explanatory sense does not also have ontological implications. A successful theory will have to posit something that is explanatorily relevant to the target phenomenon. I am saying only that the fact that the theory is about (in the explanatory sense) Xs does not have the direct implication that the theory must posit (as part of its fundamental ontology) Xs.

behave in the way they do. But a theory of the chemical properties of molecules might be a reductive theory, one whose fundamental ontology is limited to nucleons, electrons, and their electromagnetic features. It can still be true, even though the theory does not involve a fundamental ontology of molecules (and so is not about molecules in the ontological sense), that it is straightforwardly about molecules in the explanatory sense. This is because through describing the behavior of nucleons and electrons, it is able to capture all of the features of molecules we were interested in explaining. It is because of the reduction of molecules to nucleons and electrons that this is possible. So, what a theory is about in the ontological sense and what a theory is about in the explanatory sense may be distinct.

Now one might argue that a reductive theory will not provide a genuine explanation of its target unless it includes an explicit statement of the reductive relationships tying the target of explanation to the fundamental ontology of the theory. In the above case, one might insist that the reductive theory itself include a claim like:

(C) Molecules are identical to configurations of protons, neutrons, and electrons.

If so, then even in the case of a reductive theory, the explanatory sense of "what a theory is about" would impose a direct constraint on the ontology. If X is something the theory is aimed at explaining, then X would have to be at least part of the ontology of the theory, if not in the fundamental dynamical equations, then in the statements describing the reductive relationships between X and the terms in the equations. My own view is that C and claims like it are metaphysical statements that better belong to what Bell would call "the surrounding talk" of a theory (1987, p. 52). However, even if they are included as part of the reductive theory, this would still not make the target entities part of the theory's *fundamental* ontology.

In this case, even if our chemical theory is targeted at explaining molecules and we concede reference to molecules must occur as part of the theory at least in claims like C, this will still not require that molecules be part of the theory's fundamental ontology. Since molecules are constituted by the nucleons and electrons according to clause C, they are derivative ontology.

There is a question about whether this analogy with chemistry is appropriate when we are considering the status of wave function realism. One main point of contention between those holding the primitive ontology view and wave function realists is whether we can plausibly see the explanatory targets of quantum theories as reducible to the wave function. Wave function realists do typically insist that the wave function and any other elements of the high-dimensional ontology constitute what physicists regard as subatomic matter and radiation. Chapters 6 and 7 will work to spell out how this may be so.

Assuming this project succeeds, we may see how a quantum theory can explain the behavior of what is familiar (and so be about what is familiar in the explanatory sense), while not containing what is familiar in its fundamental ontology. At the same time, it is worth noting that it is not really the case that all of the proposals for a primitive ontology have been so familiar to us: matter density fields and GRW flashes hardly seem familiar entities, even if particles and particle trajectories have become so. Ultimately, it seems that the advocates of the primitive ontology view do not really keep to their insistence that the primitive ontology of a physical theory be what is antecedently familiar.[5]

Ultimately, the best defense of the primitive ontology approach is not any kind of deductive argument that quantum theories must have ontologies that meet the four constraints, but rather something more like an inference to the best explanation. This is

---

[5] Allori notes (personal communication) that such further departures from the manifest image become reasonable as one tries to accommodate relativity.

something that has been more explicit in some of Allori's work, especially her 2013 paper "On the Metaphysics of Quantum Mechanics." There, Allori makes the claim that we do not necessarily have to think that physical theories have primitive ontologies. Indeed, she concedes it might be that in some cases a primitive ontology is not available. But other things being equal, if we have an interpretation of a theory with a primitive ontology, then this should be preferred:

> The common problem to theories in which the wave function is taken as describing matter is that they are incredibly radical in an unnecessary way: if less farfetched alternatives work, why go radical? Even if we grant that there are reasons to go that way, still the theory does not seem to provide (at least at the moment) something to be compared to the explanatory [power] we just described which is available to theories with a primitive ontology. (2013b, p. 16)

The explanatory power Allori is referring to is the ability to explain the existence of macroscopic objects in a way analogous to what may be done using classical physics. It is just so much easier, she argues, to see how macroscopic objects may be constituted by a microscopic ontology in ordinary three-dimensional space or spacetime, rather than by a wave function in a high-dimensional space.

I would argue, however, that on balance, it is not true that primitive ontology interpretations are to be preferred on the basis of an inference to the best explanation. For there are theoretical virtues that favor the approach of the wave function realist. To see this, consider Table 5.1.

Those quantum theories preferred by defenders of the primitive ontology approach like Allori tend to be less ontologically parsimonious than those with only a wave function. The wave function realist need only believe fundamentally in a wave function, while one

LOCAL BEABLES 189

Table 5.1 Versions of Quantum Mechanics

Approach to Solving the Measurement Problem	State Representation	Law(s) of State Evolution
Many-worlds quantum mechanics:		
I (S_0)	Wave function	Schrödinger equation
II (S_m)*	Wave function + representation of matter density field	Schrödinger equation Matter guidance equation
Spontaneous collapse theory (GRW):		
I (GRW_0)	Wave function	Stochastic GRW equation
II (GRW_m)*	Wave function + representation of matter density field	Stochastic GRW equation Matter guidance equation
III (GRW_f)*	Wave function + representation of flashes	Stochastic GRW equation Flash equation
Bohmian mechanics*	Wave function + particle configuration	Schrödinger equation Particle guidance equation

Note: Interpretations that are favored by advocates of the primitive ontology approach are marked with an asterisk *.

subscribing to a primitive ontology view will believe fundamentally in both the wave function and a primitive ontology. Allori may defend her view on parsimony grounds by saying that although the wave function is an element of any quantum theory's fundamental ontology, it will not be an element of its primitive ontology. Indeed, Allori suggests (2013b, Section 5.3) that if the wave function is not taken to be a concrete entity like a field, then the wave function's status as abstract will not contribute to surplus ontology. However, confining the wave function to nonprimitive status does not eliminate it. It only suggests that it is not part of what constitutes the matter of the theory. It still exists and so is relevant for considerations of ontological parsimony.[6]

Second, Allori highlights the fact that theories with a primitive ontology are more conservative in that their ontologies are closer to that of both the manifest image and earlier physical theories. As she puts it:

In contrast to the case of wave function ontology, the primitive ontology approach reflects the desire to keep the scientific image closer to the classical way of understanding things. (2013a, p. 62)

Now, as I have noted, some think such conservatism has no place in the evaluation of rival physical interpretations. For example, Ladyman has argued:

Science is not under any obligation to recover familiar truths from the manifest image, only approximations of them, the reasonableness of asserting them even though they are false, or their persistence as illusions. . . . Conserving common sense is not a desideratum for physical theory or its interpretation and it may well be that "most of our ordinary beliefs about the physical world are false," . . . or at least only approximately and partially true and in other ways systematically wrong. (2010, p. 155)

Ladyman thus suggests that we do not at all need an interpretation of a physical theory to be conservative in this sense; we only need to recover why we thought earlier (perhaps mistaken) ontologies were correct.

One may easily grant with Ladyman that common sense and earlier scientific theories may have led us to have many false beliefs about the world around us. And so we should keep an open mind

---

[6] See Schaffer (2010b) for further discussion of this point, that "the proper rendering of Ockham's razor is 'Do not multiply *fundamental* entities without necessity'" (p. 313).

about the possibility of new theories revealing that the world around us is very different than we previously thought it was. But this does not mean that when weighing different interpretations against each other, it should not be considered a virtue of an interpretation that we are capable of understanding it. This may come about partly because the interpretation in question shares something in common with earlier pictures of the world we already understand. In this respect, Allori is correct that primitive ontology approaches have the virtue of maintaining a fundamental ontology closer to that of earlier (classical) theories and the manifest image.

Still, there is another sense of conservatism and with respect to this virtue, the primitive ontology approach does rather poorly. This is a matter of being faithful not to earlier scientific or quasi-scientific pictures of the world, but rather to previous interpretations of the physical theory in question. The primitive ontology approach favors ways of providing ontologies for quantum theories that are revisionary in this latter respect. These involve adding new laws and new ontology to earlier formulated quantum theories. Bohmian mechanics was a revision to orthodox quantum mechanics. GRW_m and GRW_f are revisions to the earlier GRW_0; S_m to the earlier S_0. To the extent that conservatism is a virtue that a framework for interpretation may have, there may too be a reason to remain faithful to the ontologies of earlier versions of quantum mechanics itself.

Finally, I would like to suggest that the advantage Allori claims for the primitive ontology approach, that of having a relatively simple way of explaining the macroscopic features of matter, ends up a wash. For the primitive ontologies that Allori and her collaborators take to constitute macroscopic objects do not on their own have all that is needed to ground most familiar properties (colors, textures, etc.) of macroscopic objects. Bohmian particles and GRW flashes may have spatiotemporal locations, matter density fields, variations in intensity. But to ground the existence and behavior of macroscopic objects as we know them requires more than what is contained in any primitive ontology. Additional information is needed that is contained in the wave function.[7] One may concede that the existence of a primitive ontology of matter in spacetime makes the starting point of an account of the constitution of macroscopic objects very straightforward. But this should not be viewed as a decisive reason to reject wave function realism. For, as it stands, primitive ontology interpretations also require more philosophical work, to give a full account of the character and behavior of macroscopic objects.

And this will require better understanding of the ontological status of the wave function according to the primitive ontology approach. Here, the dominant view is that the wave function is something with a nomological character. Like a law, it is something that guides and governs the primitive ontology. However, several critics point out issues with this characterization of the wave function.[8] For example, can the wave function really be nomological, that is, a law, if it itself evolves according to a dynamical equation, for example, the Schrödinger equation? Allori, Goldstein, and Zanghì give a response to these worries, speculating that in a final quantum theory, a quantum cosmology, the Schrödinger equation may be replaced with something that looks more like the Wheeler-DeWitt equation of canonical quantum gravity. In this equation, as we noted in Section 5.3, the time variable drops out, and we may see the wave function as static and unchanging, thus more like a law. This goes some way toward answering the worry, but holds the very coherence of the view hostage to one speculative proposal about quantum gravity. What is more, one wants to know what is to be said for viewing the wave function as nomological when one is interpreting the quantum theories we use today.

Another feature that makes the wave function seem less like a law is the physical contingency of its initial state. It is natural to

---

[7] This is a point David Albert has been making for some time.
[8] See especially Brown and Wallace (2005) and Belot (2012).

think there could be different physically possible worlds with different initial wave functions. However, if the wave function is nomological, then this is ruled out. To say that a contingent fact about the initial state of the universe has the status of a law is not completely unheard of today. Albert and Loewer, for example, argue that the proposition that the universe started in a low entropy state, a proposition they call the Past Hypothesis, should be understood as a law (Loewer 1996, Albert 2000). This classification of the Past Hypothesis is motivated by their preferred theory of laws, the Humean best system analysis, in which laws are regarded as those statements that appear in the description of our world that best maximizes simplicity and informational strength (Lewis 1983). However, it is not clear that this strategy may be applied to the case of the initial state of the wave function. The initial state of the wave function of our universe is extremely complex. This makes it less plausible that a statement about it would appear in the Humean's best system. Goldstein and Zanghì acknowledge this point and claim as a result that

> we should expect somehow to arrive at physics in which the universal wave function involved in that physics is in some sense simple—while presumably having a variety of other nice features as well. (2013, p. 103)

This raises an interesting issue of how one might characterize the initial wave function so that it is simple enough to occur in our best system. But we cannot tell now whether this expectation will be met.

Sometimes, Goldstein and Zanghì instead conjecture that perhaps the wave function is only quasi-nomological: it is not quite a law, but at least more like a law than a concrete, physical object (2013, Section 4). The wave function would be at least quasi-nomological because its role is to govern the primitive ontology of the theory, and, as they note, there is no "back action" in the other direction from the primitive ontology back onto the wave function. This distinguishes it from fields in other theories (for example, the electromagnetic field) that influence the behavior of the matter of those theories, but for which there is also reciprocal influence of the matter back on the fields.

Perhaps the best one can do is say that the wave function is itself a novel kind of object and should not, in the context of the primitive ontology approach, be lumped into another ontological category. This is what Maudlin has recommended (2013, 2019). This still raises questions about the status of the wave function. For example, if the wave function is not a law but is still able to affect the behavior of objects in the primitive ontology like a law, then does this mean it lacks a spacetime location? If not, then how does it affect the primitive ontology?

In this book, my wish is not to dismiss the primitive ontology approach to the interpretation of quantum theories. As I've noted a few times already, at this stage of inquiry, it is best to have more, not fewer, approaches to interpretation available. And I don't think the considerations just enumerated decisively show the wave function cannot be nomological. However, I have tried to show that there is no very compelling argument (deductive or inductive) that physical theories, including quantum theories, should be thought to have primitive ontologies. Such approaches to interpretation raise many challenging philosophical problems, and don't obviously accrue more theoretical benefits than other approaches. Thus, if we are to think that an ontology for a quantum theory must be such as to be able to constitute macroscopic objects, this isn't best motivated by claiming that quantum theories should have primitive ontologies.

5.5 Perception and the Macroscopic

We have so far seen two arguments that may be given for the first premise of the constitution objection. An ontology for quantum theories must be capable of constituting macroscopic objects because:

A. An interpretation for our best quantum theories should be empirically coherent, and to be empirically coherent, it must include an ontology of macroscopic objects,

and:

B. Quantum theories, like other physical theories, come with primitive ontologies, and this means that their ontology includes the entities that make up the macroscopic objects of our ordinary experience.

I will argue that a simpler and more compelling argument may be provided for the first premise of the constitution argument, namely, that:

C. We know macroscopic entities exist, and since our best quantum theories aim to capture a great deal of the world around us, they should include an account of the constitutive basis of these macroscopic objects, including ourselves.

Simple-minded though it is, this, I believe, is the best way to get the constitution objection going. (C) rests neither on contentious views about confirmation nor the explanatory targets of our physical theories.

(C) starts with the commonsensical claim that we can know that there exist macroscopic objects, including ourselves. We know this by our everyday perception of the macroscopic world around us. This is a claim that has been rejected by epistemological skeptics and idealists, who bring to bear sophisticated philosophical challenges involving Cartesian demons and brains in vats, but it is a claim that non-skeptical philosophers have devised equally many compelling ways of refuting (for an overview, see Hazlett 2014). Indeed the rejection of such skeptical scenarios is at least a tacit starting point of most scientific inquiry. Noting then that we can know such macroscopic objects exist, one can then argue that since it is a goal of quantum theories to provide deep accounts of much of the world around us, then these quantum theories ought to be such as to capture the existence of macroscopic objects, if not by representing them directly, then by describing the materials out of which they are constituted.

Chapters 6 and 7 accept this challenge of describing how a wave function in a high-dimensional space may come to constitute macroscopic objects including ourselves. My view is that this can be done, and so even though the first premise of the constitution argument is plausible, the argument as a whole fails. An interpretation for a quantum theory need not include the postulation of fundamental local beables.

6

The Causal Role
of Macroscopic Objects

6.1 The Macro-Object Problem

We now arrive at the final part of this book. We have already seen
how wave function realism presents us with an intriguing way
of viewing the world in light of quantum phenomena like entanglement. Wave function realists do not regard the world as ultimately constituted by a collection of objects spread out in space
or spacetime, connected by irreducible relations and able to affect
each other instantaneously across spatial distances. Instead, they
propose that what appear to be spatially separated objects are really
manifestations of a more fundamental single object, the quantum
wave function, with the appearance of unmediated interaction
across spatial separations explained by the evolution of a wave function in a higher-dimensional framework for which there fails to be
any nonlocal action. Wave function realism allows explanations for
what would otherwise be inexplicable, brute relations and action at
a distance, and thus a resolution of quantum mysteries. This is the
positive case for wave function realism.

We have also gone some way toward showing how wave function
realism may be extended to provide interpretations of quantum
theories beyond nonrelativistic quantum mechanics, in terms of
which these foundational questions are usually debated. We have
shown that wave function realism provides a general recipe for
interpretation, yielding high-dimensional, separable ontologies;
this recipe is applicable even to theories in which particle number

The World in the Wave Function. Alyssa Ney, Oxford University Press (2021). © Oxford University Press.
DOI: 10.1093/oso/9780190097714.003.0006

is indeterminate or the relevant basis is something other than the position basis.

Finally, we have addressed objections that the interpretations of quantum theories yielded by wave function realism are misguided for failing to accommodate the empirical evidence for quantum theories and facts about what our quantum theories are about. I argued that our empirical evidence for quantum theories and their purpose depends on what the correct ontological framework is. If the primitive ontology approach is correct, then the evidence for our quantum theories ultimately consists of local beables and this is what our quantum theories are fundamentally about. If wave function realism turns out to be correct, then the evidence for our quantum theories ultimately consists in facts about the wave function and this is what our quantum theories are fundamentally about. So, objections that the evidence for or targets of quantum theories must involve local beables ultimately beg the question against the wave function realist.

Now it is time to see how this higher-dimensional ontology relates to the world with which we seem to interact, how macroscopic objects, even we ourselves, might ultimately be constituted by a wave function in a high-dimensional space. Thus, these final pages concern what has been called *the macro-object problem* for wave function realism. This is the problem of showing how macroscopic objects like tables and chairs, stars and planets, measuring devices, humans, and other organisms may be constituted ultimately out of a wave function in a high-dimensional space. This is a problem that has sometimes been conflated with the measurement problem. As we have seen, the measurement problem is the problem of how quantum systems may evolve from states that appear to be indeterminate with respect to the value of some variable to states that appear to be determinate with respect to that variable. But the macro-object problem isn't fundamentally a problem about the evolution of quantum states. As we will see, whether quantum states collapse or fail to collapse, whether there are "many worlds" or only one, there will be the problem that if what is described by these quantum states is a wave function in a high-dimensional space, how this wave function could make up something like a table or a person, something that appears to possess the three (ordinary) dimensions of height, width, and depth.

In the present chapter, my goal will mostly be to canvass solutions to the macro-object problem that have been proposed and defended by other authors. I will spend the majority of time evaluating the functionalist proposals outlined (separately) by David Albert and David Wallace. My own view is that the tools of functionalism alone cannot provide an explanation for how the wave function may constitute (genuine) three-dimensional objects. In the following chapter, I present my own preferred solution to the macro-object problem, and explore how quantum theories may force us to revise some of our most basic assumptions about material constitution and ourselves.

6.2 An Initial Proposal

Given the way that states of the wave function have been introduced in earlier chapters, we can see that, at least for the nonrelativistic case, there is a rather straightforward connection between states of the wave function and configurations of a system of particles or atoms in three dimensions. And so there is at least an initial story we may tell about the connection between the wave function and macroscopic objects in an attempt to solve the macro-object problem. This starts by noting that for nonrelativistic quantum mechanics, the wave function is defined on a space with the structure of a classical configuration space in which, relative to a particular coordinatization, each point corresponds to a specific particle configuration. Given this fact, finding a solution to the macro-object problem may seem trivial. Look at the locations in the high-dimensional space where the wave function has nonzero amplitude.

Infer the existence of the corresponding three-dimensional particle configurations, and then infer a macro-ontology where these configurations involve particles appropriately clumped together.

We may note that because quantum systems are generally fields smeared out over the wave function's space, fields that fail to have their amplitude confined to one or another point-sized location in it, the matter of finding macroscopic objects (at least in a way that will manage to roughly accord with our ordinary experience) will not be quite as simple as that. For since the wave function typically takes on nonzero values at many, many points in its space, if we take the existence of a particle configuration to be entailed by the wave function's having nonzero amplitude at the corresponding point in its space, we will infer not just one configuration of particles, but rather many, many—a continuum many in fact—and so, many, many such macro-objects.

This problem would not be quite so bad if we could be confident that we would be committed only to many slightly varying particle configurations. So, for example, the wave function might have nonzero amplitude at a point corresponding to a configuration of particles that includes a collection C that is clumped up in a way to look like a chair. And then it might also have nonzero amplitude at a point very nearby corresponding to a configuration in which all the particles in C but one are clumped up in that way. And another in which in addition to the particles in C, there is one more clumped up near them. And so on. Then, according to the account we are considering, the wave function would support the existence of many chairs in the same approximate three-dimensional location, not just one. But maybe this is not so bad. After all, the problem we would face would be similar to the Problem of the Many (Unger 1980) that metaphysicians by now have many ways of addressing (e.g., van Inwagen 1990, Lewis 1993, Sider 2003).

But the wave function will generally take on nonzero values at regions corresponding to much more wildly different microscopic arrangements. So much so that if we were to follow this proposal, we would have to accept a story about what kinds of macroscopic objects exist that is not even remotely consistent with our ordinary experience. We would have to accept not just the existence of many slightly different chairs present in one approximate three-dimensional location, but also many tables and buildings and leopards and many other bizarre objects we don't have names for.

We may mollify this problem to some extent by noting that versions of quantum mechanics adequate to solving the measurement problem have built into their accounts of the dynamical evolution of the wave function the consequence that in situations corresponding to ones in which we are making observations of macroscopic objects, the wave function will not be spread over its space in such a way as to be so incompatible with our experiences. This is achieved in different ways by collapse and no-collapse approaches to solving the measurement problem. Collapse approaches work by mostly doing away with the conflicting situations. No-collapse approaches work not by doing away with the conflicting situations, but by telling a story that explains why we never experience them.[1]

First, according to spontaneous collapse approaches to quantum mechanics like GRW, individual quantum systems always have a small probability of undergoing a hit or collapse of their wave function. As we've seen, the effect of this hit may be captured mathematically by the multiplication of the initial quantum state by a Gaussian function localized around a particular point in one of the three-dimensional subspaces of the wave function's space. (Recall that each of these three-dimensional subspaces of the higher-dimensional space may be associated with one of the system's N particles.) When the wave function is in a state corresponding to the entanglement of a large number of microscopic particles, as for example will be the case in the situation in which you or I interact with a table, because of the large number ($\sim 10^{23}$) of particles involved, it

---

[1] I'll continue to focus on Everettian no-collapse solutions to the measurement problem, rather than hidden variable approaches.

is likely that the wave function will undergo a hit localizing it in at least one of these three-dimensional subspaces. And due to entanglement, this will localize the wave function in the subspaces corresponding to the locations of all of the other particles making up you or me and the particles with which we are interacting. As a result, in such situations, the wave function has much higher amplitude in some regions than it does in others. Thus, to return to our case of the many chairs, although the wave function would continue to have some nonzero amplitude in regions corresponding to particle configurations conflicting with our experience (those in which some particles are arranged in the shape of a table or a leopard), these amplitudes are very small, comparatively speaking.

Given this fact about what is entailed by collapse approaches, Albert and Loewer (1996) proposed that we adopt the following rule:

> PosR: Particle x is in region R iff the proportion of the total squared amplitude of x's wave function which is associated with points in R is greater than or equal to 1-p, where p is some real number between 0 and 1, smaller than ½ but not too small.

This proposal for reducing facts about three-dimensional particle locations to facts about the wave function is quite different from that considered above. First, we don't look at particular points in the wave function's space. Instead we consider three-dimensional subspaces of the wave function's space and look to see if there are regions in these subspaces in which the wave function has high amplitude. As noted, the collapse process doesn't completely localize the wave function in these subspaces but instead determines that it has higher amplitude in some regions rather than others. PosR then says we postulate the existence of a particle in the spatial region corresponding to the region in the three-dimensional subspace that has high amplitude.

This is how things might go for spontaneous collapse approaches. In Everettian approaches to quantum mechanics, one might worry one cannot appeal to a rule like PosR because there is no true collapse of the wave function. However, although the wave function will generally be spread out in multiple regions of its space, Everettians note that in situations in which a quantum system interacts with its environment, the quantum state will typically undergo a dynamical process of decoherence that generates what is often described as a process of "effective collapse of the wave function" onto a number of determinate states corresponding to distinct classical worlds (Zeh 1970, Zurek 1991, 2003). Although decoherence does not suppress the relative high amplitude of the wave function in locations corresponding to the existence of macroscopic objects incompatible with our experience, the process of decoherence ensures that there will not be interaction (perceptual or otherwise) between these conflicting configurations. And so the world will contain classically-behaving macroscopic objects, just more of them than we ever observe.

To see how this works, consider a single atom initially in an entangled state of x-spin:[2]

$$\psi_S = a\,|x\text{-up}>_S + b\,|x\text{-down}>_S,$$

where the subscript 'S' refers to the atomic system, and a and b are nonzero. For our purposes, we may assume a and b are real numbers. Suppose we measure this system with a Stern-Gerlach apparatus, which we may denote using the subscript 'A.'

The apparatus starts with its screen blank and may record a value for the atom's spin as "up" or "down" depending on whether a mark appears on the upper or lower part of the screen. Suppose

---

[2] This discussion closely follows Zurek (2003), p. 730. See also Wallace (2012), pp. 77–81.

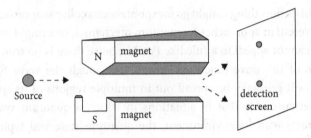

Figure 6.1. Stern-Gerlach Apparatus

the interaction between the apparatus and a system may be described as:

$$|\text{x-up}\rangle_S |\text{blank}\rangle_A \longrightarrow |\text{x-up}\rangle_S |\text{“up”}\rangle_A$$
$$|\text{x-down}\rangle_S |\text{blank}\rangle_A \longrightarrow |\text{x-down}\rangle_S |\text{“down”}\rangle_A$$

That is, the Stern-Gerlach apparatus doesn't change the spin of an atom with determinate x-spin. It only records the atom's spin with a mark on its screen. Then, by the linearity of Schrödinger evolution, we may see the joint system involving our atom in a superposition of x-spin evolving as:

$$(a\,|\text{x-up}\rangle_S + b|\text{x-down}\rangle_S)\,|\text{blank}\rangle_A \longrightarrow a|\text{x-up}\rangle_S|\text{“up”}\rangle_A$$
$$+ b|\text{x-down}\rangle_S|\text{“down”}\rangle_A$$

For a pure state, we have seen that the density matrix ρ may be written as $|\psi\rangle\langle\psi|$. So:

$$\rho_{SA} = a^2|\text{x-up}\rangle_{SS}\langle\text{x-up}||\text{“up”}\rangle_{AA}\langle\text{“up”}| + ab|\text{x-up}\rangle_{SS}\langle\text{x-down}||$$
$$\text{“up”}\rangle_{AA}\langle\text{“down”}| + ba|\text{x-down}\rangle_{SS}\langle\text{x-up}||\text{“down”}\rangle_{AA}\langle\text{“up”}|$$
$$+ b^2|\text{x-down}\rangle_{SS}\langle\text{x-down}||\text{“down”}\rangle_{AA}\langle\text{“down”}|$$

This density matrix ρ_{SA} has four terms. The second and third terms are the "off-diagonal" or "cross" terms. For general quantum states, there will be cross terms and they measure the coherence in the quantum state.

Decoherence occurs when a system interacts with its environment so as to eliminate these cross terms. Suppose the interaction with the environment works in the following way:

$$(a|\text{x-up}>_S| \text{``up''}>_A + b|\text{x-down}>_S| \text{``down''}>_A) |E_0>_E \longrightarrow a|\text{x-up}>_S|$$
$$\text{``up''}>_A|E_1>_E + b|\text{x-down}>_S| \text{``down''}>_A|E_2>_E$$

So after the interaction with the environment, the system's total wave function is:

$$\psi_{SAE} = a|\text{x-up}>_S| \text{``up''}>_A|E_1>_E + b|\text{x-down}>_S| \text{``down''}>_A|E_2>_E$$

If the states of the environment E_1 and E_2 are orthogonal (their inner product is identically 0), then it will turn out that the reduced density matrix of the joint atom-apparatus system which results when we trace out the environment will be:

$$\rho_{SA} = a^2|\text{x-up}>_{SS}<\text{x-up}|| \text{``up''}>_{AA}< \text{``up''} | + b^2|\text{x-down}>_{SS}<\text{x-down}||$$
$$\text{``down''}>_{AA}< \text{``down''} |$$

As we can see, the cross terms have gone away. Everettians take this to indicate that observers on the "up" branch of the wave function will not see interference from observers on the "down" branch. The two branches don't interact and thus evolve independently of one another. As Zeh put it, the up and down components of the total quantum state become "dynamically decoupled". There is one "world component" in which the experimentalists will observe "up" and another in which they will observe "down". And these components cannot communicate (1970, p. 75). On this Everettian model, the evolution of the wave function corresponds to many internally consistent and stable observations of a macroscopic world.

One other technical point about decoherence that is worth mentioning, as it relates to our discussion at the end of Chapter 4, is that decoherence with a system's environment generally works to single out a preferred basis. Decoherence leads to the process of einselection, and the preferred basis that gets einselected is what Zurek labels the pointer basis:

> When expanded, the [density matrix for the system+apparatus] contains terms that are off diagonal when expressed in any of the natural bases consisting of the tensor products of states in the two systems. Their disappearance as a result of tracing over the environment removes the basis ambiguity. (Zurek 2003, p. 730)

And:

> Not all quantum superpositions are treated equally by decoherence. Interaction with the environment will typically single out a preferred set of states. These pointer states remain untouched in spite of the environment, while their superpositions lose phase coherence and decohere. . . They are the preferred states of the pointer of the apparatus. They are stable, and, hence, retain a faithful record of and remain correlated with the outcome of the measurement in spite of decoherence. (2003, p. 717)

Einselection of the pointer basis is crucial for the Everettian's recovery of classicality and explanation of why we observe determinate results for measurement processes. We earlier considered a criticism of wave function realism that wave functions should not be considered fundamental for their failure to be basis-independent. But what we are seeing is that the explanation of the appearance of classically-behaving macroscopic objects from within the context of Everettian quantum mechanics relies on the process of einselection and the singling out of the pointer basis as preferred. Thus, the prioritizing of one basis appears justified, at least if we are aiming to recover a world of macroscopic objects from a fundamental Everettian quantum ontology.

Returning to the general point of this section, whether we consider collapse or no-collapse implementations of wave function realism, it appears we can find macroscopic objects in the wave function by finding subregions in the wave function's space at which the wave function has a significant nonzero amplitude, postulating the existence of particles at locations corresponding to these subregions, and building macroscopic objects out of them.

6.3 Monton's Challenge

It may at first appear that way, but unfortunately, wave function realists (including Albert and Loewer) and their critics agree that the story I've just told is inadequate. For although finding a suitable mapping between a macro-ontology and the structure of the wave function may provide perhaps a necessary condition for grounding the reality of macro-objects in the universal wave function, this is not sufficient. For the wave function's space does not itself possess any structure indicating what is the preferred way to map its points onto particle configurations. This is a point that has been argued by Brad Monton (2002):

> It could be that the x, y, and z coordinates of particle number 3 in the three-dimensional space correspond to the seventh, eighth, and ninth dimensions of the 3N-dimensional space, or it could be that they correspond to the ninth, eighth, and seventh dimensions, or it could be that they correspond to the 30th, 25th, and 240th dimensions, and so on. These different correspondences entail different ways that the N particles evolve, given a particular evolution of the objects in the 3N-dimensional space. (2002, p. 267)

Worse, there is nothing in the structure of the wave function's space to suggest its dimensions are grouped into threes, so that its points may correspond to particle configurations in a three-dimensional space rather than a space of some other dimensionality. If 3N is even, we could map states in the wave function's 3N-dimensional space onto configurations of 3N/2 particles in two-dimensional space, or 3N particles in one-dimensional space, and so on. If we want to deny that all of these other deviant ontologies exist as well, we will need to find more than a correspondence between points in the wave function's space and particle configurations in three-dimensional space.

Tim Maudlin has raised a related concern:

> The notion is that the dynamics of a very high-dimensional object in a high-dimensional space could somehow implicitly contain within it—as a purely *analytical* consequence—a description of local beables in a common low-dimensional space. This approach turns critically on what such a derivation of something isomorphic to local structure would look like, whether the derived structure deserves to be regarded as *physically* salient (rather than merely mathematically definable). (2007a, p. 3161)

Maudlin's worry is that even if there is some way to provide a mapping from features of the wave function onto the description of a low-dimensional ontology, these features may also be mapped onto additional, incompatible, nonfundamental metaphysics. But what is there to distinguish any particular mapping of the wave function onto a system of local beables as physically salient? What can a wave function realist say to make it convincing that in virtue of the existence of this one mapping but not the others, there really are these things, these low-dimensional objects, in the wave function realist's ontology?

Maudlin discusses two ways in which the wave function could ground multiple derivative ontologies. First, like Monton, he is concerned that there are differently *dimensioned* low-dimensional ontologies we may map to any particular wave function ontology:

What the wavefunction monist has is just a field evolving in this very high-dimensional space, quite unlike what we take ordinary space to be. It is true that under a certain fictive mapping, one can associate that evolution with the motion of local beables in an ordinary-looking space, and that motion corresponds to what we see. But under other perfectly well-defined fictive mappings, it will correspond to a bizarre evolution of local beables in ordinary space, or a bizarre evolution in a bizarre space, and so on. Why should one pay any more attention to one fictive mapping than another?

In addition, facts about the wave function could be used to derive facts about different kinds of microscopic local beables. Is the low-dimensional ontology that arises from the wave function one of a system of *particles* scattered over three-dimensional space? Or is it a matter field taking on various field values at three-dimensional locations? Maudlin argues that the wave function realist's fundamental metaphysics underdetermines which set of microscopic local beables there are, since each such mapping is possible. His suggestion of course is that one move to an interpretation in which local beables are fundamental:

> It is clear what would solve the problem: remove all the talk of fiction! If one believes that in addition to the wavefunction *there really is* an ordinary space that *really does* contain local beables that *really do* evolve in a specified way determined by the wavefunction, then you have something. The existence of an infinitude of other merely fictive local beables in a merely fictive space definable from the wavefunction is neither here nor there: *they* obviously play no role in the explanation of our experience. (p. 3166)

When Maudlin says one should solve the problem by saying "there really is" a three-dimensional space, it is clear that he wants us to add the low-dimensional space and objects in it as additional *fundamental* parts of one's theory.[3] But as we will see, this is not necessary.

6.4 Albert's Proposal

As I said, Albert agrees with this point about the insufficiency of finding a mapping. His response is that it is correct: there is nothing about the *synchronic* state of a wave function that makes it constitute a particular three-dimensional ontology, or a three-dimensional ontology as opposed to a two-dimensional one. For any instantaneous state of the wave function, it is of course possible to map it onto many different low-dimensional ontologies. To see the three-dimensionality in the wave function and the kind of three-dimensional ontology that arises from it, we have to look at its behavior over time. As he puts it, "The thing to keep in mind is that the production of the geometrical appearances is—at the end of the day—a matter of *dynamics*" (2013, p. 54). It is the wave function's dynamical behavior that suggests three-dimensionality.[4]

Albert's proposal is that we adopt functionalism in order to see how three-dimensional objects may be constituted out of the wave function. He calls this "functional enactment" (2013, 2015). In other words, the suggestion is that we reduce three-dimensional

---

[3] The talk of "fictive" mappings may be distracting here. What's going on is that Maudlin is responding to Albert's 1996 paper in which Albert distinguishes the fundamental, real high-dimensional space, from the illusory (hence, "fictive") three-dimensional space that arises as the result of the behavior of the wave function. Albert now holds (2015) that the emergent low-dimensional space is for its nonfundamentality no less real than the fundamental space. It is not a fiction, not illusory.

[4] Again, the talk of "appearances" might be distracting here. In Albert's work, we find two points of view: first, that the three-dimensional world is an illusion, but that its appearance may be causally explained by the dynamical behavior of the wave function; second, that the three-dimensional world is genuine and that its existence may be constitutively explained by the dynamical behavior of the wave function. In his most recent work (2015), Albert unambiguously endorses the latter position (one that he has held for some time) and so that is the view I will engage with here.

macroscopic objects to the wave function by providing a functional reduction.

Functional reductions involve two steps: first, characterizing the functional role of the entity or entities that are the target of the reduction; and second, identifying some entity or entities in the more fundamental ontology that is or are capable of playing that functional role (Kim 1998). Once this is achieved, we are justified in claiming that the target entities (here, three-dimensional objects of the kind with which we are familiar from our perceptual engagement with the world) are constituted by entities in the more fundamental ontology. It is in this way that this proposal improves on the idea that we may recover three-dimensional objects simply by mapping them onto synchronic states of the wave function. The existence of a mapping of one thing onto another doesn't suffice for the existence of a constitution relation between them. But if we can characterize *what it is for there to be* a three-dimensional object in terms of the playing of some functional role, and the wave function plays that role, then the wave function will ipso facto be capable of constituting three-dimensional objects.

A comment on space: there is a genuine question about the metaphysical status of the three-dimensional space that might arise given such a functionalization. Wave function realists typically think of the wave function's space in absolutist terms: as a substance, as the arena against which the dynamics play out, as something objects have locations in, and as a fixed and permanent background (see Albert 1996). But given this understanding of what it is to be a space, space is not the kind of thing that may be functionalized. Something that has a causal/functional analysis is not a background to dynamics, but rather a result of dynamics; it is not fixed, but evolves. And therefore, a wave function realist should not claim to show that there is an absolute three-dimensional spatial background that exists as the result of the behavior of the wave function. The only absolute space/spatial background in the wave function realist's metaphysics will be the wave function's space.

But the wave function realist need not recover three-dimensional space as an absolutist would conceive of it. The important question is whether she is able to recover macroscopic objects, where these are assumed to be objects that are three-dimensional in the sense of possessing ordinary heights, widths, and depths. Given this aim, we may allow relationalist or structuralist (Esfeld and Lam 2006) positions about the three-dimensional space that don't require for objects to genuinely be three-dimensional that there be a separate substance or background they are located in. This space may instead be constituted by a pattern of relations.

Albert's functionalism then, in my view, is best understood as positing a form of structuralism about three-dimensional space, that three-dimensional space is not a substance or background, rather a structure of (causal or dynamical) relations in something else (the wave function). If so, then we must ask first, what relations must obtain for there to be a set of entities correctly described as three-dimensional? And second, is the wave function or its parts capable of instantiating these relations?

Albert answers these questions, but first it will be useful to back up and see in outline the structure of his solution to the macro-object problem. Albert's approach to the macro-object problem involves tackling it in two stages:

Stage 1: Showing how an ontology of individual, three-dimensional *microscopic* objects may be recovered from a wave function in a high-dimensional space.

Stage 2: Showing how a three-dimensional *macro-ontology* may be recovered from this three-dimensional micro-ontology.

What is novel here is the first stage. I won't have much to say about the second stage of Albert's proposal. The idea is that once we have recovered a three-dimensional microscopic ontology of particles or fields, we may recover three-dimensional macroscopic objects in the usual way by building them out of microscopic objects by mereological composition or another functional reduction.

Now to simplify the discussion, let's work in the framework of spontaneous collapse approaches to quantum mechanics.[5] So we can assume the GRW collapse process described earlier: in the kind of situation we would ordinarily describe as a perceptual interaction with a macroscopic object, the wave function is very likely to undergo collapse, a hit that involves its bunching up with high amplitude around a location in a three-dimensional subspace of the wave function's space.

According to the account considered in the previous section (PosR), it was sufficient for there to be microscopic particles with three-dimensional locations (ordinary three-dimensional locations) if this was the case. But now, accepting Monton's challenge, we are assuming that the existence of these bunchings doesn't suffice to give us a three-dimensional ontology. So, thus far we just have these bunchings-up of the wave function in these three-dimensional subspaces.

Albert (2015, p. 130) introduces a function that takes in locations in one of these three-dimensional subspaces and spits out the amplitude of the wave function at those locations. For each subspace i, he calls this function the i-th shadow of the wave function. We may think of these shadows as matter fields that correspond to (but are not so far identified with!) individual particles in some N-particle system. So, corresponding to each GRW hit, there is a shadow in a three-dimensional subspace of the wave function's space.

---

[5] This is best for the purpose of engaging with Albert's proposal because he is skeptical that Everettian quantum mechanics provides an adequate solution to the measurement problem.

Albert argues that these shadows are capable of playing the causal/functional role of three-dimensional micro-objects. And then it is "the relatively stable coagulations of subsets of these shadows" that play the role of tables and chairs. Let's now look at Albert's account of the causal/functional role of three-dimensional micro-objects and how shadows in the wave function may be capable of playing this role.

One way of giving a description of how a kind of physical system behaves over time is by specifying its Hamiltonian. Again, this may be thought of (for some kinds of systems) as a sum of kinetic and potential energy terms. Albert notes that the following is a way of writing down a Hamiltonian $\hat{H}$ for a classical, three-dimensional system of N particles:

$$\hat{H}_1 = \sum_i m_i \left(\left(\frac{dx_i}{dt} \right)^2 + \left(\frac{dy_i}{dt} \right)^2 + \left(\frac{dz_i}{dt} \right)^2 \right)$$
$$+ \sum_i \sum_j V_{ij} \left((x_i - x_j), (y_i - y_j), (z_i - z_j) \right)$$

The i and j indices range over the particles in the system, and we can look at this Hamiltonian as summarizing how the behavior of the system over time depends on the masses and velocities (along the x, y, and z dimensions) of the individual particles in the system, as well as the spatial distances between the particles, again in these three dimensions. Objects' later positions depend on how fast they're moving and in what direction, and whether there are other particles moving toward them and potentially colliding into them. This is Albert's functional characterization of what it is (roughly) for a system of objects to be three-dimensional: for their behavior to be described by a classical Hamiltonian like this, that there are objects whose behavior over time depends on changes in position and interparticle distances in three dimensions (Albert 1996, 2013, 2015).

And his claim is that the wave function shadows are capable of playing this functional role. This condition is satisfied when the hits occur so that we may have a stable coordinatization of the wave function's space in 3N dimensions and the behavior of the wave function is describable roughly by the Hamiltonian:

$$\hat{H}_2 = \sum_i m_i \left(\left(\frac{dx_{3i-2}}{dt} \right)^2 + \left(\frac{dx_{3i-1}}{dt} \right)^2 + \left(\frac{dx_{3i}}{dt} \right)^2 \right)$$
$$+ \sum_i \sum_j V_{ij} \left(\left(x_{3i-2} - x_{3j-2} \right), \left(x_{3i-1} - x_{3j-1} \right), \left(x_{3i} - x_{3j} \right) \right).$$

Here, the indices i and j range over the shadows and a location $(x_{(3i-2)}, x_{(3i-1)}, x_{3i})$ picks out the location of the center of mass of the ith shadow. Then, according to Albert:

> It will follow that the clean and literal and unadorned picture of the GRW theory that we are considering here is going to accommodate relatively stable three-dimensional coagulations of subsets of these shadows in the shapes of tables and chairs and baseballs and observers, and that the effects that these shadow-tables and shadow-chairs and shadow-baseballs and shadow-observers have on one another ... are going to be the ones that we ordinarily associate with the tables and chairs and baseballs and observers of our everyday experience. (2015, pp. 137–138)

If the wave function evolves in accordance with such a Hamiltonian, the three coordinates of each shadow's subspace (3i–2, 3i–1, 3i) will play the same role as the x, y, and z coordinates represented in $\hat{H}_1$. And thus the shadows will constitute genuine three-dimensional objects.

6.5 Troubles with Functionalism

Let's grant the assumption that the GRW hits may occur in such a way that we could approximately describe the wave function using a Hamiltonian like $\hat{H}_2$. Still, I want to argue that it is incorrect to say that this suffices to generate a three-dimensional micro-ontology. The glaring difference between what is described by $\hat{H}_1$ and $\hat{H}_2$ for all their similarity in form is that in the first case, the behavior of the system depends on particle velocities and distances in three dimensions: x, y, and z. In the second case, the behavior of the system depends on velocities and positions (though not distances) in 3N dimensions: x_1 through x_{3N}. When we focus on any individual shadow, we may describe its evolution in a three-dimensional subspace of the wave function's space. But once we begin to talk about the relationship between velocities of individual shadows and inter-shadow "distances" (the differences referred to in the potential energy term of $\hat{H}_2$), we see that the shadows are not related in any common three-dimensional space. There is no common three-dimensional space for interparticle interactions, let alone interparticle distances. Each shadow lives in its own three-dimensional subspace. For example, when i = 1 and j = 2, note the fact that the potential energy V in the first Hamiltonian partly depends on the value $x_i - x_j$ or $x_1 - x_2$ which is simply the distance (in the x–dimension) between particles 1 and 2. In the second Hamiltonian, V depends on the value $x_{(3i-2)} - x_{(3j-2)}$ or $x_1 - x_4$ where this refers to the numerical difference between the first coordinate of the center of mass of the first shadow in its three-dimensional subspace and the first coordinate of the center of mass of the second shadow in its own (different) three-dimensional subspace. This isn't a distance; it's merely a difference between values. And so, there is no common three-dimensional space in which the shadows may "coagulate" and make up tables and chairs.

Albert's shadows do not possess distance relations to one another in the wave function's space, but could we perhaps instead say that the difference in values described by the second Hamiltonian, even if it isn't a fundamental distance relationship, plays the role of a distance relation, because it is this difference in values that makes for differences in shadows' later states? Of course, if we are assuming that $\hat{H}_2$ approximately describes some quantum system, then this means these differences in values are relevant to the later behavior of our system. But this doesn't thereby entail that this difference in values plays the functional role of an interparticle distance. This would be to vastly bleach out the functional characterization of what it is to be an interparticle distance. Something is an interparticle distance not merely if it is a value upon which the later state of a system depends.

Perhaps Albert is thinking that these differences play the role of distances not merely because they are relevant to later states of the particles, but also because they mediate interactions between the shadows. But it is not clear what grounds there are for saying there is an interaction between shadows in this picture. Hits occur spontaneously in the framework and so are not caused. In entangled states, the manner of localization of the hit in one subspace does bear a modal correlation with the manner of localization of the others, but as I argued in Chapter 3, there are good reasons not to think of this as a causal interaction.[6]

At this stage, all that seems left to motivate the claim that the second Hamiltonian describes common three-dimensional locations and distances is its formal similarity to $\hat{H}_1$. But vastly different kinds of systems may all be described using common mathematical notation. Thus, this fact about formal/mathematical similarity alone does not suffice to entail any metaphysical conclusions. Even granting that $\hat{H}_1$ can describe the causal/functional role characteristic of a three-dimensional system of particles,

---

[6] And recall that if hits are causing other hits, this challenges the locality the wave function realist wants to have in the fundamental high-dimensional image.

being representable by a Hamiltonian formally like $\hat{H}_1$ is not sufficient for playing this role.

Later in his (2015), Albert considers a different strategy that may seem to address the concern I've just raised. (Though he doesn't present this as an improvement on the first, he presents both as adequate ways of capturing a macro-ontology in the wave function.) The strategy involves defining the "3D compression of the wave function" and then proposing that various compressions can play the role of macroscopic objects. A 3D compression of the wave function is a function defined as the direct sum of the N shadows of the GRW wave function. Each shadow is defined on a distinct three-dimensional subspace. But now we ignore this fact and simply sum the amplitudes. To do this, of course, we will have to decide on a common coordinatization. But assume we do that. Then we can read facts about where this field is clumped up in three-dimensional space at different times off of this three-dimensional representation. This is very similar to an interpretation of quantum mechanics that was considered by Schrödinger (1927, cf. Allori et al. 2011).

The crucial thing to note is that compression is a mathematical operation, a summing. There is no grounds for speculating that a new physical process takes place that corresponds to this operation of summing substates. So it is not that a new three-dimensional space has resulted somehow. Instead, there are still rather only all of these individual three-dimensional subspaces, but superimposing them mentally, we get something that seems like a three-dimensional microscopic world that could functionally enact tables. But what we wanted was a three-dimensional arrangement, not a way of representing what is in reality a 3N-dimensional arrangement in three-dimensional terms.

My conclusion is that even assuming GRW, there just doesn't seem to be anything available in the fundamental ontology of the wave function realist to get something to play the causal-functional role of three-dimensional objects, of things that move toward and away from one another with various velocities in three dimensions and according to the distance between them.

This concludes my main objection to Albert's way of using functionalism to establish the derivative reality of three-dimensional objects. However, even if the approach was successful in the cases for which the second Hamiltonian applies, there is an additional concern with Albert's approach, namely, that it ties the wave function's enactment of a three-dimensional ontology to its approximation of classical behavior. This raises the question of how the wave function realist might recover those macro-systems exhibiting nonclassical, that is, distinctively quantum behavior. Quantum effects have been known to occur for some time in surprisingly large microscopic systems and more recent work reveals quantum behavior in macroscopic systems as well (e.g., Eibenberger et. al. 2013). Departure from classical behavior doesn't seem to track departure from three-dimensionality in the appearances. One might object that large-scale quantum behavior has only been demonstrated in experimental contexts with classical-behaving detectors. But these experiments are thought to reveal quantum behavior that would be present too in the absence of such detection devices. In any event, it would be nice to at least have available a way to recover three-dimensionality that does not assume classicality.

6.6 The Decoherence Strategy

Those favoring Everettian approaches to solving the measurement problem have pursued another strategy for recovering macroscopic objects from the wave function. As we discussed in Chapter 4, Everettians like David Wallace and Chris Timpson are skeptical about wave function realism as a framework for interpreting quantum theories beyond nonrelativistic quantum mechanics. But

they disagree with those who have objected to wave function realism by claiming that the wave function is in principle incapable of
constituting macroscopic objects. A priori, there are no conceptual
obstacles, in their view, to seeing three-dimensional macroscopic
objects as arising as nonfundamental elements of a wave function
ontology:

> 3-dimensional features will emerge as a consequence of the dy-
> namics (in large part due to the process of decoherence): the form
> of the Hamiltonian will ensure that the wave-function decomposes
> into components evolving autonomously according to quasi-
> classical laws. That is, each of the decohering components will cor-
> respond to a system of well-localized (in 3-space) wavepackets for
> macroscopic degrees of freedom which will evolve according to
> approximately classical laws displaying the familiar 3-dimensional
> symmetries, for all that they are played out on a higher-dimensional
> space. 3-dimensional quasi-classical structures are thereby
> recovered. (2010, p. 705)

Timpson and Wallace see the wave function as capable of generating
local beables as the result of a contingent but natural dynamical process (decoherence). The decoherence process allows localized systems
to persist in the wave function's space, systems that behave as if they
existed in a classical, three-dimensional world in virtue of exhibiting
the symmetries of three-dimensional systems.

As we saw in Section 6.2, decoherence obtains when a physical
system interacts with its environment so as to produce an entangled
composite system. This system can then be represented by a sum of
states in the pointer basis such that, when we factor out the degrees of
freedom of the environment, we obtain a reduced density matrix that
eliminates the off-diagonal terms corresponding to superpositions
of determinate states. What an appeal to decoherence achieves is the
elimination of the chance that an experimenter will detect a superposition of incompatible outcomes.

Decoherence is certainly an important tool for the wave function realist in the Everettian context. What it allows her is a way of representing quantum states in terms of states with determinate (classical) outcomes. And the decoherence process leads to einselection, thereby picking out a three-dimensional position basis, the pointer basis.[7] However, although decoherence picks out a privileged basis, which it is natural to refer to as the pointer basis, this does not suffice to establish a privileged correspondence between points in the wave function's space and three-dimensional configurations, thus addressing Monton's challenge. Einselection chooses a basis, and thus may privilege a kind of wave function representation. But it does not make it the case that there is a preferred mapping of these wave function states onto a low-dimensional ontology.

I am not trying to argue here that decoherence plays no role in connecting what we observe to features of the wave function or even that those who invoke decoherence are confused about the role it plays. It is just one piece of the puzzle. In Wallace's work (see especially 2004), he notes that quantum theory forces us to adjust our conception of the relationship between microscopic and macroscopic phenomena in order to understand how our world may be constituted out of a quantum ontology. When classical mechanics was the leading candidate for a fundamental physics with its ontology of particles in three-dimensional space, we could see macroscopic objects as built up out of these classical particles like houses are made up of bricks. But since the wave function doesn't come in little bits[8] each of which might correspond to the tiny, microscopic parts of a chair or a table or a cat, we must revise our view of the relationship between the fundamental physical ontology and the macroscopic. Wallace's own suggestion (2004, 2010, 2012) is that

---

[7] Whether it does this is more controversial once we extend the discussion to the relativistic domain.

[8] It doesn't come in bits that are little from the point of view of its own space and plausibly identical with tables and chairs.

we adopt a more flexible picture of this relationship; we may view classical objects as emerging from the quantum state in virtue of their existence as *patterns* in that state.[9] If the higher-dimensional ontology is structured in the right way, then this is sufficient for there to be a three-dimensional macroscopic object in any genuine sense one could ask for. The three-dimensional object just is a pattern in the quantum state (2012, p. 60).

In the context of answering the constitution objection, however, we must ask whether this appeal to structure and patterns will satisfy anyone who wants an understanding of how a field in a high-dimensional space could constitute low-dimensional macro-objects. What does it mean for the quantum state or wave function to be "structured" in the right way for a chair to be a pattern in it? The passage from Wallace and Timpson above suggests an answer. Parts of the wave function are patterned or structured in the right way in virtue of evolving "according to approximately classical laws displaying the familiar three-dimensional symmetries." This is another appeal to functionalism. The decohering components of the wave function may come to constitute genuine three-dimensional objects in virtue of behaving in a way compatible with familiar three-dimensional symmetries.

But here I worry we are back to the same problem that Albert's proposal faced. In what sense are the parts of the wave function obeying three-dimensional (as opposed to only 3N-dimensional) symmetries? This must be fleshed out in a way that requires more than just that the evolution of the wave function may be represented by a Hamiltonian similar to $\hat{H}_1$. But in quantum mechanics, the presence of a symmetry is analytically tied to the condition that the generator of that symmetry commutes with the system's Hamiltonian. Thus, I am not sure that there is a way of putting flesh on the idea that a certain system obeys certain symmetries that would both (a) describe a behavior the wave function or its parts

[9] Wallace finds inspiration in the work of Daniel Dennett (1991).

may genuinely instantiate, and (b) not come to a merely formal condition of being describable by a certain kind of mathematical representation. As we've seen the latter is too weak.

Finally I think we have to be careful about importing the functionalist's language of patterns to this case.[10] We can list many cases where the pattern concept seems to unproblematically apply: checks and zigzags are patterns in the colored threads making up a sweater, hurricanes and tornados are patterns of movement of regions of high temperature and pressure, a dance is a pattern of movement and gestures. But what is common to each case is that the less fundamental entity depends on the spatial or spatiotemporal arrangement of bits of the more fundamental ontology. The pattern exists precisely because there is a spatiotemporal arrangement in the more fundamental ontology. But the demand on the wave function realist who wants to solve the macro-object problem in order to respond to the constitution objection is to show how there can be a spatiotemporal arrangement in the first place.

As we will see in Chapter 7, I do believe symmetries are an important element of the story of why we should think certain systems are three-dimensional, rather than, say two-dimensional. However, there is more work to do to demonstrate that the wave function itself is enacting genuine three-dimensional symmetries, rather than merely simulating them.

6.7 From Simulation to Constitution

In order to reject the second premise of the constitution objection, it will be essential for the wave function realist to find a way of demonstrating that low-dimensional macroscopic objects may be constituted out of something like the wave function. In light of

---

[10] This is not to say that the appeal to patterns is implausible in all cases, for example, as part of a functionalist solution to the mind-body problem.

Monton's challenge, it is a reasonable first step to try to find some parts of the wave function that are capable of playing the functional role of low-dimensional objects. However, as I have argued in this chapter, such strategies as have so far been tried fail to be convincing. It is possible to arrive at representations that bear similarities to representations of low-dimensional objects, so that we may say a wave function may simulate a low-dimensional object. But what we need is something more, a way of seeing the wave function as making up (in some sense) low-dimensional objects. For this, we need to establish the existence of a constitutive relationship. This will be my goal in Chapter 7.

7

Finding the Macroworld

7.1 A Constitutive Explanation in Two Stages

I have argued that previous strategies fail to convincingly recover a genuine ontology of three-dimensional macroscopic objects from the behavior of the wave function. Thus, they fail to convincingly recover objects of the kind we seem to interact with in our everyday experiences. Albert's functionalist strategy for recovering macro-objects requires the wave function to play a role it can't, as a high-dimensional object. And Wallace's proposal, to the extent that it does not similarly require that the wave function be a three-dimensional object, one in which three-dimensional patterns may inhere, relies on too weak a criterion for genuine three-dimensionality. Functionalist strategies for recovering a three-dimensional world from the behavior of the wave function thus seem to have a challenging path to navigate. One has to, in the first place, make sure that the requirements for three-dimensionality are requirements the wave function or its parts are actually capable of satisfying. But one additionally has to make sure that the requirements are not so weak that they may be satisfied by what are clearly not three-dimensional objects.

My aim in this chapter is to find a different way for the wave function realist to convincingly establish the reality of local beables. This will involve filling out in more detail what it is about the behavior of the wave function that makes it yield a derivative low-dimensional ontology. If one likes, one may regard this as a way of developing the symmetry-based proposal of Wallace and Timpson. However, I would add two caveats. First, I am interested in an approach that

The World in the Wave Function. Alyssa Ney, Oxford University Press (2021). © Oxford University Press.
DOI: 10.1093/oso/9780190097714.003.0007

does not require for three-dimensionality that the wave function or its parts approximate classical, i.e. nonquantum behavior. Second, as we will see, I don't regard what I will propose as a functional reduction of three-dimensionality to a wave function ontology. There is a role for functionalism, in my view. But this comes only at a later stage once a three-dimensional ontology of microscopic objects is already established.

My account then takes part in two stages. These are distinct from the two stages of Albert's proposal. Recall, Albert first recovered individual three-dimensional microscopic objects from the wave function and then built macroscopic objects out of them. Instead, I suggest we start by recovering whole three-dimensional configurations of micro-objects from the wave function, and then moving from these large configurations to find our macroscopic objects. The reason for this shift is straightforward: the recovery of three-dimensionality comes from direct relations between states of the wave function as a whole and total three-dimensional micro-configurations. And so that is where we will begin. First, however, I would like to clarify the role in my account of a notion that has come to be central in metaphysical discussions of the constitution of macroscopic objects, since it has been thought by some to provide an easy way out of these difficulties concerning the recovery of derivative macroscopic objects from a more fundamental wave function ontology. This is the notion of metaphysical grounding.

7.2 The Role of Grounding

In her "The Structure of a Quantum World," Jill North (2013) proposes that we see macroscopic, three-dimensional objects as related to the wave function by grounding relations:

> I think that there is a way of making sense of the idea that ordinary space is nonfundamental yet real, in the same way that ordinary

objects, the special sciences, and so on, are nonfundamental yet real. A *grounding relation* captures the way that the wave function's space is fundamental and ultimately responsible for ordinary space, while at the same time allowing for the reality of ordinary space. (2013, p. 198)

As I will explain, this proposal on its own is not sufficient to do the work the wave function realist needs to do here.

Grounding frameworks have been developed in different ways in the work, among others, of Kit Fine (2001), Gideon Rosen (2010), and Jonathan Schaffer (2009). Characterized in broad strokes, grounding introduces a way of conceptualizing the sense in which a more fundamental ontology may provide the constitutive or more broadly metaphysical basis for another, derivative ontology or set of facts. To provide a full grounding explanation for some fact leaves no further unanswered questions about in virtue of what that fact obtains.

> If the truth that P is grounded in other truths, then they *account* for its truth; P's being the case holds *in virtue of* the other truths' being the case. (Fine 2001, p. 15)

For example, when we say that the existence of a certain house is grounded in there being a collection of bricks arranged in a particular kind of configuration, we are providing a metaphysical or constitutive explanation of the existence of the house by showing in virtue of which more basic ontological constituents and facts about them the house exists.

For Fine and Rosen, not all cases of grounding or metaphysical explanations are constitutive explanations; they are sometimes by contrast eliminative. To use Fine's example, a mathematical nominalist, one who denies the reality of numbers and other mathematical objects, may argue that what appear to be facts about a Platonic realm of numbers[1] are grounded in some more fundamental facts

---

[1] For example, facts such as that there are prime numbers greater than one million.

about concrete objects, such as mathematicians and the proofs they make. The existence of the mathematicians and their proofs explain in virtue of what these mathematical facts obtain, according to the nominalist. Such grounding explanations—while clearly metaphysical explanations stating what there is in reality in virtue of which these mathematical facts obtain—are different from the grounding explanation of the house. These are eliminative, not constitutive explanations. Instead of showing us what mathematical objects like numbers are constituted from, they show us how facts seeming to be about numbers may still hold even though numbers are not real.

It is because grounding explanations may be of such a variety of different types, ranging from the clearly constitutive to the clearly eliminative, that a mere appeal to grounding will not suffice to solve the macro-object problem for wave function realism. For the wave function realist's goal in addressing this problem is to make it convincing that there may genuinely be low-dimensional macroscopic objects of the kind we seem to interact with, even though our world is fundamentally a wave function in a high-dimensional space.[2] But, as Jessica Wilson has correctly noted (2014, pp. 244–248), to simply say that one fact grounds another leaves completely open what the relationship between these facts is, and indeed even whether the derivative facts describe any realm of genuine objects. She labels this the metaphysical underdetermination problem for theories of grounding.[3]

My own view, unlike Wilson's, is that we needn't thereby discard the notion of grounding. Rather we should seek to provide constitutive explanations that would explain exactly how it is that facts about the wave function may ground facts about the existence of

---

[2] This isn't to say that seeing the world as constituted out of a wave function may not license us to revise some of our beliefs about the features of these macroscopic objects. We know from the history of scientific reductions that these always involve at least some revisions to our concepts of what is reduced. See Hooker (1981) and Bickle (1999).

[3] For further discussion, see Ney (2016) and (2020b).

genuine macroscopic objects, not merely their appearances. And this is what I aim to do in what follows.

For metaphysical aficionados, it may be worth noting that what I have said up until now may seem to apply more directly to the grounding framework proposed by Fine and Rosen, rather than that of Schaffer. For Schaffer, grounding is a metaphysical relation between entities, some of which are seen as relatively fundamental (the grounds) and others that are relatively derivative (the grounded). Because grounding is a relation between entities, there is no possibility for Schaffer of what is grounded failing to exist. And so one might object to what I've just said by noting that if one interprets North as positing a Schafferian grounding relation between the wave function and macroscopic material objects, then there is no chance of the macroscopic being eliminated or in some sense not fully real, nor of there being some sort of metaphysical underdetermination in the sense of Wilson.

Indeed, I do think this is the sort of grounding framework North has in mind. A similar proposal—that we see macroscopic objects as grounded in the wave function—has been made by Ismael and Schaffer (2020). But note that even if the problem of elimination is avoided, simply positing grounding relations still does not adequately address the macro-object problem since until we understand *how* the wave function may constitute macroscopic objects, there will still be an explanatory gap. And then critics of wave function realism, those advocating the constitution objection, will still complain that wave function realists can't solve the macro-object problem.

Now, some may be less concerned than others about explanatory gaps. Schaffer himself argues that actually there are many more explanatory gaps than we usually think there are, but these shouldn't trouble us. Even in those paradigm cases for which we seem to have an understanding of how microscopic objects may come together to constitute macroscopic ones (e.g., those in which two hydrogen atoms and an oxygen atom may come together to form a water molecule), there are still brute grounding relations that remain unexplained:

> *Explanatory gaps are everywhere.* There is no transparent rationale in any of the standard connections, even from the H, H, and O atoms to the H_2O molecule, since it is not transparent that the H, H, and O atoms compose anything, much less something with the nature of an H_2O molecule. Correlatively, I claim that nothing of moment follows from such gaps, so long as they are bridged by principles of metaphysical grounding. The connections in question are bridged by substantive mereological principles concerning the existence and nature of wholes, which mediate metaphysical explanations just as laws of nature mediate causal explanations. In a slogan: *Grounding bridges gaps.* (2017, p. 2)

If one likes, one can of course postulate grounding relations that link atoms and molecules, that link the wave function with macroscopic reality.[4] Indeed, in the philosophy of mind literature, many self-professed physicalists have learned to live with the epistemic possibility that explanatory gaps will always remain between what we know about physical reality and the facts about consciousness.

But I am not one of those philosophers who thinks we should accept the existence of explanatory gaps. Rather, I think that the fruit of collaborations between philosophers, cognitive scientists, and neuroscientists gives us reason to be optimistic that we can bridge explanatory gaps even in the case of linking our understanding of consciousness and of the physical world. The desire to remove explanatory gaps and, in Schaffer's words, *make transparent* the

---

[4] As discussed in Chapter 2, Schaffer himself is not a wave function realist but a priority monist adopting a low-dimensional fundamental spacetime. But his point about grounding and explanatory gaps is more general, and he is explicit that this is a metaontological point of view the wave function realist should accept as well (2017, p. 16).

relationship between physical states and consciousness is exactly what motivates the most exciting work on the mind-body problem and elevates its standards for successful resolution. But even were the acceptance of explanatory gaps to be tolerated in that case or even in the case of water and H_2O, this strategy would not carry over to be useful for the wave function realist. For in these other cases, there is no controversy about whether the two ontologies linked by the putative grounding relations are genuine. Few doubt the existence of water molecules or the more fundamental atoms in which they are purportedly grounded. Few doubt the existence of conscious states or the more fundamental brain states in which they are purportedly grounded. But the macro-object problem for wave function realism is a problem that stems from those who do very much question the reality of a more fundamental wave function (*qua* field in high-dimensional space). And so making transparent this constitutive relationship is quite important for making the ontological claim the wave function realist wants to make compelling in the first place. This is not a challenge we face in other contexts in which we at first find explanatory gaps. So, even if we posit a grounding relation between the wave function and macroscopic objects, this must only be a first step. There remains the significant challenge of spelling out how the wave function may constitute the existence of (genuine) macroscopic objects.

7.3 Recovering Three-Dimensionality Using Symmetries

I will now begin to outline my proposal for addressing the first stage of a solution to the macro-object problem for wave function realism: that is, recovering an ontology of three-dimensional objects from the wave function. I will begin at the same place that Albert began: noting, in response to Monton's challenge, that the presence of a mapping of synchronic states of the wave function onto three-dimensional particle configurations does not suffice to establish that states of the wave function may constitute a three-dimensional, derivative reality. I agree with Albert that the way to find three-dimensional objects in the wave function is to look at the latter's dynamical behavior rather than focusing on its state at a single time. However, I will suggest a different way of showing the physical salience of particular three-dimensional ontologies, one that I would not call a functional reduction since the strategy is not to start by identifying the causal/functional role of a three-dimensional ontology and then find something in the wave function ontology that plays that role.

My strategy is instead to start with some facts about invariances between quantum states (states of the wave function) and then to make use of a principle that invariances are generally indicative of the presence of symmetries. I show how in certain cases, symmetries are revealed if we presume the existence of a genuine three-dimensional ontology, but are not similarly revealed if we presume the existence of a two- or other-dimensional ontology, a deviant three-dimensional ontology, or the fundamental high-dimensional ontology alone. And so, in such cases, the link between dynamical invariances and symmetries gives us reason to see the wave function as constituting a particular three-dimensional ontology.

I'll explain what I have in mind using an example of one such invariance, that of the permutation group, because it is simple. The principle of permutation invariance says that permuting the positions of particles makes no difference to the dynamical behavior of a total system. So, for example, a two-particle system in which one electron is on the left and another is on the right is identical, from the point of view of the laws, to one in which the second is on the left and the first is on the right.

Consider a situation in which the universal wave function is defined on a 12-dimensional space, and consider two points in this space that we may represent as follows:

P1: $(-1, 1, 0, 1, 1, 0, -1, -1, 0, 1, -1, 0)$
P2: $(1, 1, 0, -1, 1, 0, -1, -1, 0, 1, -1, 0)$.

Now consider two states: one in which the amplitude of the wave function is clumped around P1, and another in which the amplitude of the wave function is clumped around P2. Here is a picture of what this corresponds to in terms of three-dimensional particle arrangements. (The z-axis is not pictured.) Think of the numbers as names of the individual particles (Figure 7.1).

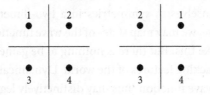

Figure 7.1. Three-Dimensional Image

One holding that symmetries represent redundancies in a theory would say that these represent the same state. But let's be neutral about that for now. Simply suppose that the laws make no distinction between the state in which the amplitude of the wave function is clumped around P1 and the state in which the amplitude of the wave function is clumped around P2. Given a three-dimensional interpretation, the fact that we do not observe differences in the behavior of the corresponding systems under the laws is reflected in the form of a permutation symmetry between particle configurations. The only difference between the two three-dimensional particle configurations is that particles 1 and 2 are permuted.

Now consider two two-dimensional interpretations of our two wave function states. In the first case, two of the particles map onto the same point (Figure 7.2).

In the two-dimensional case, we don't need particle labels. These configurations manifestly are not symmetric. What we see are not mere differences in which particles are located where, but where

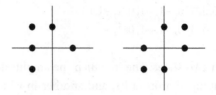

Figure 7.2. Two-Dimensional Image

there are particles and their relative distances. So the dynamical invariances among these states of the wave function do not manifest themselves in symmetries in a two-dimensional representation. Sure, we may map states of the wave function onto a representation like this. But there is nothing to be gained by doing so in terms of tracking features of the world. Dynamical invariance in states of the wave function thus may distinctively legitimize three-dimensionality. It also picks out particular configurations, distinguishing them from the deviant ones Monton considers.[5]

There are several concerns one might have with this proposal. First, note that nothing I have said rules out that the invariances might also manifest themselves in interesting symmetries we find under some other interpretations that are not three-dimensional, or under some alternative three-dimensional interpretations. This, however, does not undermine the proposal.[6] Suppose the dynamical invariances in the quantum state are equally well reflected in permutation symmetries of four particles in a three-dimensional space and of two particles in a six-dimensional space. Then we could say that both interpretations are legitimate as they both reveal the symmetries tracked by dynamical invariances. However, this multi-layered metaphysics shouldn't concern us. After all, these aren't additional *fundamental* metaphysics that would be providing

---

[5] Showing this is left as an exercise for the reader.

[6] Note this concern could equally be raised for Albert's proposal. Nothing logically rules out a system exhibiting behavior approximately described by multiple Hamiltonians.

us with conflicting accounts. Rather these accounts are all compatible.[7] They would be complementary ways of viewing what the wave function realist believes is objectively there. And it is arguably quite unlikely that we would get the explosion of deviant ontologies Monton rightfully points out are consequences of the mere mapping approach.

One might be concerned about whether we have been assuming the right way of tracking invariances in the quantum state—by letting there be brute, unexplained invariances at the fundamental level in the first place that would be revealed in a lower-dimensional interpretation. Some have argued that what these invariances reveal is that our way of viewing the space the wave function inhabits is not quite right. Rather, where the laws do not recognize a difference between certain states of the wave function, we should believe that really there is no difference between these states.

For example, Leinaas and Myrheim (1977) argue that instead of the 3N-dimensional space we have been discussing, we should rather view the wave function's space as a reduced state space, one that results from identifying states related by what correspond to permutation of particle indices under a three-dimensional interpretation.[8] If the wave function instead evolves in this reduced space, then there will not be any such invariances to worry about in the fundamental metaphysics.[9] Taking on this view of the wave function's space as a reduced state space, my argument would change. It is not that some three-dimensional interpretations are physically salient because they have symmetries that track invariances between the behavior of distinct fundamental states. Instead, these three-dimensional interpretations would be physically salient because

---

[7] See also Ney (2015).

[8] See French and Rickles (2003) for an endorsement of this view.

[9] It is worth pointing out that the move to this reduced state space is not completely straightforward. As Leinaas and Myrheim point out (1977, p. 5), on their way of doing this, points that correspond to two or more particles coinciding in three-dimensional locations are singularities in the reduced state space.

they have symmetries that track something else, aspects of the interesting structure of the wave function's space.

Eddy Keming Chen (2017) and Kelvin McQueen (personal communication) have raised the concern that the strategy I propose here on behalf of the wave function realist to recover three-dimensional appearances may be problematic, in that it may undermine wave function realism rather than support it. For if three-dimensional particle configurations are required to reveal what would otherwise be brute invariances in the laws, then, Chen and McQueen suggest, this means we should adopt an interpretation of quantum mechanics with a three-dimensional rather than a 3N-dimensional ontology.

In response, note first that wave function realists do not generally dispute the reality of three-dimensional ontologies, only their fundamentality. And the argument I have presented in this section does nothing to undermine the arguments in favor of the fundamentality of the higher-dimensional wave function metaphysics. One such argument was presented in Chapter 2. My preferred argument, based on the avoidance of fundamental nonseparability and nonlocality was presented in Chapter 3. If a critic wants to undermine wave function realism, she must find a way to rebut these arguments. The argument that symmetries make physically salient certain three-dimensional ontologies shows only that three-dimensionality is a real aspect of our world, that it tracks objective structure, not that it is fundamental. So it does nothing to undermine wave function realism.

To highlight one difference between this account of the manifestation of three-dimensionality and Albert's, recall that Albert's proposal works by speculating that the wave function of a world like ours behaves so that objects' trajectories depend on three-dimensional interparticle distances and velocities. As I pointed out above, it is an advantage of an account that it not require classicality—that it accommodate the three-dimensionality even of systems whose behavior over time results from distinctively quantum effects. The invariances appealed to in the account above are robust features of quantum systems that do not require classical (i.e., nonquantum) assumptions.

At the same time, I should be clear how the position I have elaborated in the present section is compatible with certain aspects of Albert's account. What I have done is provided an account of how, if one is a wave function realist, one may also legitimately claim the (nonfundamental) truth of a three-dimensional representation. Some three-dimensional particle representations reveal symmetries that track fundamental dynamical invariances. And this answers Monton's challenge by saying what distinguishes some low-dimensional ontologies as metaphysically privileged over others. But to say all of this is not to deny that the wave function may also, under certain conditions, evolve in a way that is approximately describable using a classical-looking Hamiltonian like $\hat{H}_2$. As we will see, this will play some role in the complete story of how macroscopic objects of the kind we observe are constituted by the wave function. It just does not help in this first stage of recovering three-dimensionality.

I have argued that Albert's proposal does not succeed because the functional role of a three-dimensional particle configuration that Albert specifies is one that the wave function cannot play. And the point that $\hat{H}_2$ bears a formal similarity to $\hat{H}_1$ is not sufficient for a system evolving according to $\hat{H}_2$ to be a three-dimensional system. But I am not trying to improve on Albert's functionalist strategy by specifying a better functional role for three-dimensionality, one the wave function can play. The present discussion evades the question of the correct functional analysis of three-dimensionality altogether. It seems to me that systems may be three-dimensional, even though there is no underlying invariant behavior or the disposition for it (cf. French 2014, Chapter 9); some systems may just be brutely three-dimensional. And so I am not confident that the causal/functional role of being a three-dimensional system is that there be another entity (or class of entities) that exhibits such invariant behavior. Anyway, I am much more confident that the story I have told is the right way to begin in addressing the macro-object problem than I would be in claiming that invariances adequately capture the causal/functional role of three-dimensionality. And so I won't try to shoehorn this account into a functionalist package.

7.4 Parts and Wholes

This account of the metaphysical relationship between the wave function and macroscopic objects involves multiple stages. My goal in the previous section was to answer Monton's challenge by showing in virtue of what there might exist a particular three-dimensional ontology as opposed to another, three-dimensional ontology, or a two-dimensional ontology, or some other deviant ontology. But in doing so, I did not demonstrate in what sense if at all a high-dimensional field like the wave function could be the right kind of thing to *constitute* a three-dimensional ontology of particles or atoms or macroscopic objects. To do that, and close the explanatory gap between the wave function and facts about macroscopic objects, to solve the macro-object problem and make it convincing that a wave function in a high-dimensional space could come to constitute the existence of a world like the one we seem to observe around us, one must specify the metaphysical relationship between the wave function and three-dimensional objects like particles or atoms or tables and chairs. The goal of this section is to provide a framework for addressing this latter issue.

Let's start simply by considering a situation in which a wave function has nonzero amplitude at only one point in its space. This will not happen in a world like ours, but we may still ask, conceptually, what would one say in a situation like that. Given the standard top-down way of representing locations in the wave function's space, by analogy with the configuration space of classical mechanics, one may argue for a correspondence with a three-dimensional state in which a system has many particles, and each particle possesses a determinate and precise location. Given the argument in the previous section, and assuming facts about invariances between this state and others, facts that are revealed in symmetries in the lower-dimensional representation, we may conclude that this is more than a mere correspondence. It is a physically salient, metaphysically privileged correspondence because the lower-dimensional representation reveals objective structure in what is fundamental.

In light of this point, we are justified in granting the truth or approximate truth of the statement that there are some particles positioned at determinate locations in three-dimensional space. However, this by itself does not demonstrate how these particles (or the three-dimensional macro-objects they might compose) may themselves be constituted out of the wave function.[10] In other words, the appeal to symmetries, though it satisfactorily answers Monton's challenge, is not enough to solve the macro-object problem. To do so, we must, as a first step, clarify the relationship between the wave function and low-dimensional particle configurations.

According to the wave function realist, only the wave function and the entities of its higher-dimensional space are fundamental, and so for low-dimensional particles to be real, they must somehow be constituted out of the wave function. A natural place to start in explicating this constitution relation, in this simple case in which all of the wave function's amplitude is clustered on a single point,

---

[10] Various contemporary theorists of metaphysical fundamentality make this point that even if facts or propositions may be grounded or made true by a fundamental ontology, this does not entail that the entities these facts or propositions appear to quantify over are thereby constituted out of the fundamental ontology. See Heil (2013) for discussion of this point in the context of a truthmaking framework of fundamentality and Fine (2001) for discussion in the context of a metaphysical grounding framework. Contemporary metaphysicians generally reject the earlier view of Quine (1948) that quantificational facts directly reflect matters of ontology.

is with the hypothesis that the metaphysical relation obtaining between these many particles and the wave function is a mereological relation. The wave function is the fundamental whole; the particles its derivative parts. This picture recalls Schaffer's priority monism described in Chapter 2. Recall priority monism is the view according to which wholes are ontologically prior to their parts (in contrast to the more common view he terms "priority pluralism" according to which parts are ontologically prior to wholes).

Some may be concerned about viewing the relationship between the wave function and microscopic particles as one of whole and parts due to the fact that parts and their wholes seem typically to be located in a common spatial framework. For example, when bricks (the parts) mereologically compose a house (the whole), the bricks all possess locations at spatial regions that are proper subsets of the total spatial location of the house.

However, this is not a general requirement for mereological relations to apply to a system of parts and wholes (as stated explicitly by Varzi 2016).[11] Indeed, one might think that abstract entities not located in space or spacetime at all may exhibit mereological relations. For example, 'egalité' is part of the national motto of France. What is essential to mereology is that parthood is a partial ordering relation (it is reflexive, transitive, and anti-symmetric) and in terms of this and other logical notions, one may define notions of proper parthood (x is a proper part of $y =_{df} x$ is a part of y and $x \neq y$) and overlap (x overlaps $y =_{df} \exists z$ (z is a part of x and z is a part of y). Not any proper ordering is a parthood relation, but one may include additional principles (e.g., Supplementation: If x is a proper part of y, then $\exists z[z$ is a part of y and it's not the case that z overlaps

***

[11] This is one of the main issues explored in the 2014 volume *Mereology and Location*, edited by Shieva Kleinschmidt. Ned Markosian does argue in his contribution to that volume that [x is a part of y] be understood as (analytically) reducible to [x is located at a subregion of the region y is located at] (Markosian 2014). However, if one wants a general account of mereology that can apply to a variety of revisionary metaphysics, one ought to reject such an analysis.

x]) in order to further distinguish the parthood relation from other proper orderings (see again Varzi 2016). None of these principles require the assumption that parts and whole exist in some common low-dimensional framework.

7.5 Partial Instantiation

So, one may argue in the simple case I considered that the relation between the wave function and some microscopic particles is one of whole to parts. But what should one say more generally, allowing that in more realistic situations, the wave function will be spread with nonzero amplitude over more than one point in its space? In these more realistic scenarios, can one maintain again that the relationship between the wave function and some microscopic particles is one of whole to parts?

One reason to think that the answer is "yes" is that the wave function can in principle evolve into and out of the more special situations just considered. Through that evolution, there will be interesting differences in the wave function's state corresponding to facts about where those particles are. But these differences do not amount to changes in the mereological relationship between the wave function and those particles. An object's parts may move around without thereby ceasing to be its parts.[12]

One might be concerned about situations in which particle number is not conserved or not determinate. What about situations in which the state of the wave function evolves in such a way that the number of particles changes? This poses no conceptual obstacles. It is very common that objects gain and lose parts. What about situations in which the number of particles the wave function decomposes into is indeterminate? Analogous cases are

[12] That objects' relative locations don't determine which objects they compose is established by the arguments in van Inwagen (1990).

familiar from philosophical discussions of the Problem of the Many (Unger 1980). We may always ask which precise particles compose my hand or that table or the state of California. In recent years, metaphysicians have constructed rigorous conceptual frameworks to handle discussions of metaphysical indeterminacy (Barnes 2010, Barnes and Williams 2011, Wilson 2013). There are no conceptual barriers to there being actual situations in which it is indeterminate which parts an entity has.

Allowing this, we may now turn back to consider the simpler cases in which particle number is both conserved and determinate, though the wave function is spread with nonzero amplitude over multiple locations in its space. In those situations in which the amplitude of the wave function fails to be confined to a point, there will fail to be any determinate fact about the locations of the N particles. And where there is no determinate fact about the locations of the particles, it is a consequence that it is also not straightforward to say which if any macroscopic objects there are. We will get to this issue about the existence of macroscopic objects in a moment.

But first, once we consider situations in which the wave function has nonzero amplitude at multiple locations in its space, it becomes natural to think of the spatial configuration of the particles as a property that may be attributed to them. Then, one can say that multiple configurations of the N particles will be instantiated by the wave function, each to a degree equal to the amplitude of the wave function at that point squared. This is what I will refer to as *partial instantiation*: entities, in this case, collections of particles, may instantiate properties not only completely (to degree 1), but also partially (to some degree between 0 and 1). Typically, in contemporary philosophical discussions, we are not used to thinking of properties as instantiated only to a degree. Rather we tend to think of a property's being instantiated as an all-or-nothing matter. Entity e is either F or not F. It cannot be F only to a degree. To reject this would be to reject classical logic.

But consider how implausibly this principle applies to specific cases. For example, why should we say that Socrates must either be wise or not wise? Why isn't it more plausible to say that Socrates only partially instantiates wisdom, that he instantiates it only to a degree? Once we recognize the vagueness that pervades our concepts and their application in realistic situations, we realize that of course Socrates only instantiates wisdom to a degree. Nobody is fully wise. This was one of Plato's chief insights: nothing in the world of material objects ever perfectly has a property or lacks it:

> My good fellow, is there any one of these many fair and honorable things that will not sometimes appear ugly and base? And of the just things, that will not seem unjust? And of the pious things, that will not seem impious?
>
> No, it is inevitable, he said, that they would appear to be both beautiful in a way and ugly, and so with all the other things you asked about. (*Republic* V.479 a–b)

And so everything that is wise is always also at least a little bit not-wise. The suggestion is to make this degree-talk a bit more precise and allow that alternative particle configurations may be instantiated to degrees corresponding to the amplitude-squared of the wave function at the points corresponding to these configurations.

The account presented up until now answers the question of how there may genuinely be three-dimensional objects constituted out of the wave function. The answer is that three-dimensional particles are related mereologically to the whole that is the wave function. Although these particles usually do not have determinate locations, they may instantiate multiple locations to various degrees.

Although she does not appeal to partial instantiation, this view is very much compatible with Wilson's (2013) account of metaphysical (including quantum) indeterminacy. Wilson explicates indeterminacy in terms of a system's possessing a determinable property while not possessing a unique determinate of that determinable (either by possessing no corresponding determinate or possessing more than one).[13] In the account I am offering, the locations of the N particles are indeterminate because the particles instantiate multiple determinate locations (each to a degree).

Several pressing questions follow. Why do we not observe these particles to have multiple locations? Why do particles instead seem to have determinate locations? How are these particles with indeterminate locations related to macroscopic objects? Answering these questions requires a consideration of how the wave function evolves over time. It is here (again) where a consideration of the wave function's dynamics, and hence its Hamiltonian, must be considered.

The answer to why do we not observe these particles to have multiple locations is that we do. As was noted already, interference effects continue to be demonstrated in systems with greater and greater numbers of constituent particles. This then raises the question of why particle locations don't generally appear indeterminate. We don't observe tables as partly in one location and partly in many others. We observe things in determinate locations. Here we must reflect on the fact that the question has shifted. We do not ever directly observe particles. We directly observe macroscopic objects like tables. The view here is not that the locations of tables are indeterminate or only partially instantiated. The view is that the locations of *particles* are generally indeterminate and partially instantiated. And so the question is how particles that partially instantiate multiple locations may come to constitute tables, measuring devices, and other macroscopic objects with determinate locations.

We may return to the point that the wave function will evolve in such a way as to have higher amplitude at some regions of its space

---

[13] An example of such metaphysical indeterminacy would be an object being red, while failing to be any determinate shade of red.

rather than others. And this will entail, given the story that has just been told, that some approximate particle configurations will be instantiated to higher degrees than others. Let's consider in particular the situations one would like to describe as ones in which an observer is making an observation of a macroscopic system. It is in situations like these that the collapse dynamics of GRW will imply that the wave function will become peaked in many of the three-dimensional subspaces of the wave function's space. And thus, on the current interpretation, many ($\sim10^{23}$) particles will instantiate certain approximate locations to a very high degree.[14] These particles may then possess highly determinate (though of course not completely determinate) locations. Depending on the persistence of these highly determinate localizations and that their relationship with others may be modeled by an approximately classical Hamiltonian like that appealed to by Albert, we may talk about the stable existence of macroscopic objects, including observers themselves. Note the different role quasi-classical dynamics is playing in this account versus Albert's. The dynamics is not generating the existence of three-dimensional particle configurations. Rather the three-dimensional particle configurations are already there. They are there because the particles themselves are parts of the wave function, and these instantiate spatial configurations as determined by the amplitude-squared of the wave function at various high-dimensional locations. The role of the quasi-classical dynamics is rather to generate the existence of stable *macroscopic* objects.

The Everettian will tell a different story about the existence of macroscopic objects. Here the wave function does not collapse, but in situations we would like to describe as ones in which an observer is interacting with a macroscopic system, the evolution of the wave function is generally thought to evolve according to

---

[14] I am now talking about approximate locations since the wave function will become peaked in the form of a Gaussian function around certain regions of each three-dimensional subspace. This corresponds to the high-degree-instantiation of a localized cluster of precise three-dimensional positions.

a process of decoherence. In this case, the wave function will not evolve in such a way that we may speak of many ($\sim 10^{23}$) particles each having a highly determinate approximate location. But there will be localized clusters of high amplitude wave function in such a way that we may speak of many ($\sim 10^{23}$) particles each having several fairly determinate locations, where these locations are clustered together in a way that is compatible with various observations we may make. Because of decoherence, these separate parts of the wave function will not interact with one another. For example, in the Schrödinger's cat set-up, the particles will instantiate to a fairly high degree an approximate configuration shaped like an observer opening a box and seeing a live cat (and an unbroken flask of poison, etc.) and will also instantiate to a fairly high degree an approximate configuration shaped like an observer opening a box and seeing a dead cat (and a broken flask, and poison spilled out, etc.). In this way, assuming the wave function continues to evolve in these separated regions of its space according to quasi-classical dynamics, the N particles may be truly said to constitute inter alia both a live cat and a dead cat. Thus, the wave function realist may use part of what is now a well-worked out and familiar story of the generation of multiple macroscopic worlds from a wave function obeying Schrödinger dynamics and exhibiting decoherence.[15] Again, the obeying of local quasi-classical dynamics (in this case brought on by a globally decoherent process) is only part of the story of how macroscopic entities may genuinely exist in the metaphysics of the wave function realist. We start from the existence of three-dimensional particle configurations. These may then constitute (in this case, multiple, seemingly incompatible) macroscopic objects as they stably instantiate to a fairly high degree stable approximate positions and quasi-classical trajectories.

---

[15] For a recent overview, see Wallace (2012).

7.6 Tables, Chairs, and the Rest

As a final step, let's begin to digest what lessons this solution to the macro-object problem may have for our understanding of ourselves and the objects around us, assuming the wave function realist provides a compelling account of the fundamental reality underlying the low-dimensional appearances.

First, it is interesting to think that the world around us that appears to be spread out only in three or four dimensions of space and time, is really, more fundamentally a radically different kind of world, a field spread out over a higher-dimensional space.

Also, according to this account, macroscopic objects are abstractions from a larger reality and so not as sharply separate from their surroundings in the way our naïve picture represents them as being. All objects are in a sense aspects of the one. And this may bring with it positive associations with earlier proposals in Western (idealist) and Eastern (Buddhist) philosophical traditions. But one shouldn't draw from this the conclusion that you and I and the stars and planets are nonetheless not still distinct entities, distinct ways of abstracting from the one, more fundamental whole. To say that we are fundamentally parts of a one is distinct from saying we are one simpliciter. And so whatever ethical conclusions one would like to draw from this sort of holism, one must acknowledge this fact.

Are there any other interesting metaphysical consequences of this proposal? My claim that the wave function instantiates a wide variety of properties to various degrees may seem to have the consequence that we too and the objects around us may instantiate a wide variety of properties to various degrees. I may to a very small degree exist in many places at once, be scattered all over the globe and indeed the cosmos. And due to the variety of properties the wave function may instantiate, I may to a very tiny degree be a doctor (a real doctor), a police officer, an astronaut, perhaps even a mermaid, or to use Bertrand Russell's example, a poached egg.

Unfortunately, none of these exciting consequences follow from this proposal. Although the wave function may instantiate very many radically different particle configurations to very tiny degrees, configurations in which particles are arranged to constitute folks in many ways like you or me with interesting career paths, or fins instead of legs, there is no justification for saying that these are configurations in which you or I also exist, just having interesting other occupations or features. For on this account, objects are abstractions from the wave function that arise when there are particle collections instantiating certain properties to various degrees. Your particles may instantiate interesting other properties and so to a small degree instantiate other kinds of interesting worlds containing interesting other kinds of objects. But from that it does not follow that you do. You are not identical to these particles or any configuration of them. And so, sorry to say, we are not even to a small degree mermaids, nor mermen.

7.7 Finding the World in the Wave Function

So we can find our world of tables and chairs, stars and planets, people and measuring devices in the wave function, although we need not rule out that there may be other worlds as well.

The story I have told proceeds in several stages. First, the answer to Monton's challenge of what privileges a three-dimensional particle representation of the world is that this representation serves to reveal symmetries deriving from invariances in the quantum state, while other representations do not. This answer to Monton's challenge does not thereby solve the macro-object problem, as it does not explain how it may be that the wave function may come to constitute low-dimensional objects, but it plays an important role in making physically salient our three-dimensional image of the world as tracking ontological structure that is genuinely there.

The solution to the macro-object problem then comes in two stages, by first noting that there is no conceptual barrier to understanding the relationship between the wave function and microscopic particles as one of whole to part, and second, by noting that these microscopic particles do not have determinate positions, but only instantiate these positions partly, to some degree. Once we have particles instantiating relatively determinate positions, then it is possible to use the usual functionalist strategies to recover ordinary macroscopic objects. However, note that in this account, functionalism plays no role in recovering the three-dimensional ontology from the higher-dimensional one. It only serves to recover material objects, once their lower-dimensional, microscopic counterparts are already recovered. The relationship between the wave function and particles is one of whole to part, but no part of the wave function in its space plays the role on its own of a three-dimensional object. Three-dimensionality arises due to the behavior over time of the wave function as a whole.

Postscript: An Incredulous Stare

The goal of this book has been to better understand the picture that results when one takes seriously the idea that what quantum theories are telling us is that what appear to be spatially separated objects connected by irreducible relations of quantum entanglement and able to affect each other instantaneously across spatial distances are really manifestations of a deeper, high-dimensional reality in which spatial nonseparability and nonlocal influence dissolve away. One straightforward way of putting flesh on this idea, one informed by the kind of mathematical representations used in quantum theories, takes the world to be fundamentally constituted by a field, the quantum wave function, which inhabits a high-dimensional space. The kind of high-dimensional space one should take this field to occupy depends on the kind of physical system one is interested in representing, or in the case one wants to represent the entire quantum world, the kind of quantum theory one takes to best represent this world at this stage of scientific development. Against those who would argue for in principle barriers to understanding the world of quantum theories, there are no insurmountable conceptual difficulties that arise when one tries to make sense of this proposal, even beyond the context of nonrelativistic quantum mechanics.

The question of how best to see the macroscopic objects of our experience as constructed out of the wave function is a critical question for the wave function realist, and this is why I have devoted several chapters to addressing it. I have raised concerns about the functionalist proposals that until now have been seen as the best on offer. But it is my view that the wave function can provide an ontological foundation for a world of macroscopic objects roughly of the kind we seem to observe and otherwise interact with, and metaphysical innovations like priority monism, partial instantiation, and frameworks for capturing metaphysical indeterminacy provide the wave function realist with the conceptual tools she needs to show how.

However much wave function realism offers an intuitive and simple framework worth developing and exploring as one among many possible routes to understanding the reality underlying the phenomenon of quantum entanglement, I understand that there will likely remain a persistent resistance to the proposal that will ultimately be unshaken by the arguments presented in the previous chapters. In many ways, this resistance is characteristic of responses to radical ontological proposals. When the philosopher David Lewis proposed his modal realism, the view that our world is just one among a continuous infinity of possible worlds realizing all of the myriad ways things could be, each concrete and spatiotemporally disconnected from the others, he acknowledged that one of the most common and stubborn responses to his proposal was what he called the incredulous stare:

> I once complained that my modal realism met with many incredulous stares, but few argued objections. . . . The arguments were soon forthcoming. We have considered several of them. I think they have been adequately countered. They lead at worst to standoffs. The incredulous stares remain. They remain unanswerable. But they remain inconclusive. (1986b, p. 133)

As with Lewis's proposal, any proposal like wave function realism will be met with incredulous stares. I myself find the ontologies that result from the framework difficult to accept at a gut level. But then, this is at the same time what makes them so interesting.

It was, I believe, this kind of gut incredulity that made Schrödinger so quickly discard the suggestion that his wave function could underlie reality, and accept Lorentz's objection that the

3N-dimensional space it was supposed to inhabit could not be a physical space. But the work of philosophers of physics recounted and engaged with here have (I hope) given us both a better idea of how to make sense of this proposal, as well as a clearer account of the real challenges the proposal faces, challenges that move beyond mere blanket rejection based on sheer incredulity. As technologies develop and we begin to make more use of the surprising fact of quantum entanglement in our lives, we may hope that having one more clear and intelligible way of making sense of this phenomenon may be useful. What is intelligible, even true, may not be easy to believe.

References

Albert, David Z. 1992. *Quantum Mechanics and Experience*. Cambridge, MA: Harvard University Press.

Albert, David Z. 1996. Elementary Quantum Metaphysics. In *Bohmian Mechanics and Quantum Theory: An Appraisal*. J.T. Cushing, A. Fine, and S. Goldstein, eds. Dordrecht: Kluwer, 277–284.

Albert, David Z. 2000. *Time and Chance*. Cambridge, MA: Harvard University Press.

Albert, David Z. 2013. Wave Function Realism. *The Wave Function: Essays in the Metaphysics of Quantum Mechanics*. Oxford: Oxford University Press.

Albert, David Z. 2015. *After Physics*. Cambridge, MA: Harvard University Press.

Albert, David Z. Manuscript. How to Teach Quantum Mechanics. http://philsci-archive.pitt.edu/15584/.

Albert, David Z. and Barry Loewer. 1996. Tails of Schrödinger's Cat. In *Perspectives on Quantum Reality*. R. Clifton, ed. Dordrecht: Kluwer, 81–92.

Albert, David Z. and Lev Vaidman. 1989. On a Theory of the Collapse of the Wave Function. In *Bell's Theorem: Quantum Theory and Conceptions of the Universe. Fundamental Theories of Physics*. M. Kafatos, ed. vol. 37. Dordrecht: Springer, 1–6.

Allori, Valia. 2013a. Primitive Ontology and the Structure of Fundamental Physical Theories. *The Wave Function: Essays in the Metaphysics of Quantum Mechanics*. Oxford: Oxford University Press, 58–75.

Allori, Valia. 2013b. On the Metaphysics of Quantum Mechanics. *Precis de la Philosophie de la Physique*. S. Lebihan, ed. Vuibert, 116–151.

Allori, Valia. forthcoming. A New Argument for the Nomological Interpretation of the Wave Function: The Galilean Group and the Classical Limit of Nonrelativistic Quantum Mechanics. *International Studies in the Philosophy of Science*.

Allori, Valia, Sheldon Goldstein, Roderich Tumulka, and Nino Zanghì. 2008. On the Common Structure of Bohmian Mechanics and the Ghirardi-Rimini-Weber Theory. *British Journal for the Philosophy of Science*. 59: 353–389.

Allori, Valia, Sheldon Goldstein, Roderich Tumulka, and Nino Zanghì. 2011. Many-Worlds and Schrödinger's First Quantum Theory. *British Journal for the Philosophy of Science*. 62: 1–27.

Ariew, Roger. ed. 2000. *G.W. Leibniz and Samuel Clarke Correspondence*. Indianapolis: Hackett.

Aspect, Alain, Philippe Grangier, and Gérard Roger. 1981. Experimental Tests of Realistic Local Theories via Bell's Theorem. *Physical Review Letters*. 47(7): 460–463.

Bacciagaluppi, Guido and Antony Valentini, eds. 2009. *Quantum Theory at the Crossroads: Reconsidering the 1927 Solvay Conference*. Cambridge: Cambridge University Press.

Baez, John. 1999. An Introduction to Spin Foam Models of Quantum Gravity and BF Theory. https://arxiv.org/abs/gr-qc/9905087.

Barnes, Elizabeth. 2010. Ontic Vagueness: A Guide for the Perplexed. *Noûs*. 44(4): 601–627.

Barnes, Elizabeth and J. Robert Williams. 2011. A Theory of Metaphysical Indeterminacy. *Oxford Studies in Metaphysics*. 6: 103–148.

Barrett, Jeffrey A. 2001. *The Quantum Mechanics of Minds and Worlds*. Oxford: Oxford University Press.

Barrett, Jeffrey A. and Peter Byrne, eds. 2012. *The Everett Interpretation of Quantum Mechanics: Collected Works 1955–1980*. Princeton, NJ: Princeton University Press.

Becker, Adam. 2018. *What Is Real? The Unfinished Quest for the Meaning of Quantum Physics*. New York: Basic Books.

Bell, John. 1964. On the Einstein-Podolsky-Rosen Paradox. *Physics*. 1: 195.

Bell, John. 1976. A Theory of Local Beables. *Epistemological Letters*. 9: 11–24.

Bell, John. 1981. Bertlmann's Socks and the Nature of Reality. *Journal de Physique Colloques*. 42: 41–62.

Bell, John. 1981. Quantum Mechanics for Cosmologists. In *Quantum Gravity 2*. C. Isham, R. Penrose, and D. Sciama, eds. Oxford: Clarendon, 611–637.

Bell, John 1982. On the Impossible Pilot Wave. *Foundations of Physics*. 12: 989–999.

Bell, John. 1987. Are There Quantum Jumps? In *Schrödinger: Centenary Celebration of a Polymath*. C.W. Kilmister, ed. Cambridge: Cambridge University Press, 41–52.

Bell, John. 1989. Against 'Measurement.' In *62 Years of Uncertainty: Erice*. New York: Plenum.

Belot, Gordon. 2012. Quantum States for Primitive Ontologists: A Case Study. *European Journal for the Philosophy of Science*. 2(1): 67–83.

Benatti, Fabio, Giancarlo Ghirardi, and Renata Grassi. 1995. Describing the Macroscopic World: Closing the Circle within the Dynamical Reduction Program. *Foundations of Physics*. 25: 5–38.

Bickle, John. 1999. *Psychoneural Reduction: The New Wave*. Cambridge, MA: MIT Press.

Bohm, David. 1951. *Quantum Theory*. Englewood Cliffs, NJ: Prentice-Hall.

Bohm, David. 1952. A Suggested Interpretation of Quantum Theory in Terms of "Hidden Variables." *Physical Review*. 89: 166–193.

Bohm, David and Basil Hiley. 1993. *The Undivided Universe: An Ontological Interpretation of Quantum Theory*. London: Routledge.

Bohr, Niels. 1928. The Quantum Postulate and the Recent Development of Atomic Theory. *Nature.* 121: 580–590.

Born, Max. 1964. The Statistical Interpretation of Quantum Mechanics (1954). In *Nobel Lectures, Physics 1942–1962.* J. B. Birks, ed. Amsterdam: Elsevier, 256–267.

Born, Max and Albert Einstein. 2005. *The Born-Einstein Letters.* London: Palgrave-Macmillan.

Brown, Harvey and David Wallace. 2005. Solving the Measurement Problem: De Broglie-Bohm Loses Out to Everett. *Foundations of Physics.* 35: 517–540.

Calosi, Claudio. forthcoming. Quantum Monism: An Assessment. *Philosophical Studies.*

Cao, ChunJun, Sean M. Carroll, and Spyridon Michalakis. 2017. Space from Hilbert Space: Recovering Geometry from Bulk Entanglement. *Physical Review D.* 95(2): 024031.

Carnap, Rudolf. 1947. Empiricism, Semantics, and Ontology. In *Meaning and Necessity.* Chicago: University of Chicago Press.

Carroll, Sean. 2019. *Something Deeply Hidden: Quantum Worlds and the Emergence of Spacetime.* New York: Dutton.

Carroll, Sean M. and Ashmeet Singh. 2019. Mad-Dog Everettianism: Quantum Mechanics at Its Most Minimal. In *What Is "Fundamental"?* A. Aguirre, B. Foster, and Z. Meerali, eds. Cham: Springer, 95–104.

Cartwright, Nancy. 1999. *The Dappled World: A Study of the Boundaries of Science.* Cambridge: Cambridge University Press.

Chalmers, David. 2017. The Virtual and the Real. *Disputatio.* 9(46): 309–352.

Chen, Eddy Keming. 2017. Our Fundamental Physics Space: An Essay on the Metaphysics of the Wave Function. *Journal of Philosophy.* 114(7): 333–365.

Conroy, Christina. 2011. The Relative Facts Interpretation and Everett's Note Added in Proof. *Studies in History and Philosophy of Science Part B: Studies in History and Philosophy of Modern Physics.* 43(2): 112–120.

Dennett, Daniel C. 1991. Real Patterns. *Journal of Philosophy.* 88(1): 27–51.

Dewitt, Bryce. 1970. Quantum Mechanics and Reality. *Physics Today.* 23(9): 30–40.

Dürr, Detlef, Sheldon Goldstein, and Nino Zanghì. 1992. Quantum Equilibrium and the Origin of Absolute Uncertainty. *Journal of Statistical Physics.* 67:1–75.

Eibenberger, Sandra, et al. 2013. Matter-wave Interference with Particles Selected from a Molecular Library with Masses Exceeding 10,000 amu. *Physical Chemistry Chemical Physics.* 15: 14696–14700.

Einstein, Albert. 1948. Quantum Mechanics and Reality. *Dialectica.* 2: 320–324.

Einstein, Albert, Boris Podolsky, and Nathan Rosen. 1935. Can Quantum-Mechanical Description of Physical Reality Be Considered Complete? *Physical Review.* 74: 777–780.

Emery, Nina. 2017. Against Radical Quantum Ontologies. *Philosophy and Phenomenological Research*. 95(3): 564–591.

Esfeld, Michael. 2004. Quantum Entanglement and a Metaphysics of Relations. *Studies in History and Philosophy of Science Part B*. 35(4): 601–617.

Esfeld, Michael and Vincent Lam. 2006. Moderate Structural Realism about Space-time. *Synthese*. 160(1): 27–46.

Everett, Hugh III. 1957. "Relative State" Formulation of Quantum Mechanics. PhD dissertation. Princeton University, Princeton, NJ.

Fine, Arthur. 1996. *The Shaky Game*. Chicago: University of Chicago Press.

Fine, Kit. 2001. The Question of Realism. *Philosophers' Imprint*. 1: 1–30.

Forrest, Peter. 1988. *Quantum Metaphysics*. Oxford: Blackwell.

Fraser, James D. 2020. The Real Problem with Perturbative Quantum Field Theory. *British Journal for the Philosophy of Science*. 71(2): 391–413.

French, Steven. 2013. Whither Wave Function Realism? In Ney and Albert 2013.

French, Steven. 2014. *The Structure of the World*. Oxford: Oxford University Press.

French, Steven and Dean Rickles. 2003. Understanding Permutation Symmetry. In *Symmetries in Physics*. K. Brading and E. Castellani, eds. Cambridge: Cambridge University Press, 212–238.

Friedman, Jonathan R., Vijay Patel, W. Chen, S. K. Tolpygo, and J. E. Lukens. 2000. Quantum Superposition of Distinct Macroscopic States. *Nature*. 406: 43–46.

Ghirardi, Giancarlo, Philip Pearle, and Alberto Rimini. 1990. Markov Processes in Hilbert Space and Continuous Spontaneous Localization of Systems of Identical Particles. *Physical Review A*. 42(1): 78–89.

Ghirardi, Giancarlo, Alberto Rimini, and Tullio Weber. 1986. Unified Dynamics for Microscopic and Macroscopic Systems. *Physical Review D*. 34: 470–471.

Goldstein, Sheldon and Nino Zanghì. 2013. Reality and the Role of the Wave Function in Quantum Theory. In *The Wave Function: Essays in the Metaphysics of Quantum Mechanics*. A. Ney and D. Z. Albert, eds. Oxford: Oxford University Press, 91–109.

Griffiths, David J. 2005. *Introduction to Quantum Mechanics*, 2nd ed. New York: Prentice Hall.

Hazlett, Allan. 2014. *A Critical Introduction to Skepticism*. London: Bloomsbury.

Healey, Richard. 2002. Can Physics Coherently Deny the Reality of Time? *Royal Institute of Philosophy Supplement* 5: 293.

Heil, John. 2012. *The Universe as We Find It*. Oxford: Oxford University Press.

Heisenberg, Werner. 1927. Uber den anschaulichen Inhalt der quanten-theoretischen Kinematik und Mechanik. *Zeitschrift thfür Physik*. 43(3–4): 172–198. Translation at https://ntrs.nasa.gov/archive/nasa/casi.ntrs.nasa.gov/19840008978.pdf.

Henson, Joe. 2013. Non-Separability Does Not Relieve the Problem of Bell's Theorem. *Foundations of Physics*. 43(8): 1008–1038.

Hooker, C. A. 1981. Towards a General Theory of Reduction. Part I: Historical and Scientific Setting. *Dialogue*. 20(1): 38–59.

Howard, Don. 1985. Einstein on Locality and Separability. *Studies in History and Philosophy of Science*. 16(3): 171–201.

Howard, Don. 1989. Holism, Separability, and the Metaphysical Implications of the Bell Experiments. In *Philosophical Consequences of Quantum Theory: Reflections on Bell's Theorem*. J.T. Cushing and E. McMullin, eds. Notre Dame, IN: University of Notre Dame Press, 224–253.

Hubert, Mario and Davide Romano. 2018. The Wave-Function Is a Multi-Field. *European Journal for the Philosophy of Science*. 8(3): 521–537.

Huggett, Nick and Christian Wüthrich. 2013. Emergent Spacetime and Empirical (In)coherence. *Studies in the History and Philosophy of Modern Physics*. 44: 276–285.

Ismael, Jenann and Jonathan Schaffer. 2016. Quantum Holism: Nonseparability as Common Ground. *Synthese*. doi: 10.1007/s11229-016-1201-2.

Kim, Jaegwon. 1984. Concepts of Supervenience. *Philosophy and Phenomenological Research*. 45: 153–176.

Kim, Jaegwon. 1998. *Mind in a Physical World*. Cambridge: MIT Press.

Kleinschmidt, Shieva, ed. 2014. *Mereology and Location*. Oxford: Oxford University Press.

Kochen, S. and E. P. Specker. 1967. The Problem of Hidden Variables in Quantum Mechanics. *Journal of Mathematics and Mechanics*. 17: 59–87.

Ladyman, James. 1998. What Is Structural Realism? *Studies in History and Philosophy of Science Part A*. 29(3): 409–424.

Ladyman, James. 2010. Commentary: Reply to Hawthorne: Physics Before Metaphysics. In *Many Worlds? Everett, Quantum Theory and Reality*. S. Saunders, J. Barrett, A. Kent, and D. Wallace, eds. Oxford: Oxford University Press, 154–160.

Ladyman, James and Don Ross. 2007. *Every Thing Must Go*. Oxford: Oxford University Press.

Laudan, Larry. 1981. A Confutation of Convergent Realism. *Philosophy of Science*. 48(1): 19–49.

Leinaas, J. M. and J. Myrheim. 1977. On the Theory of Identical Particles. *Nuovo Cimento*. 37B(1): 1–23.

Lewis, David. 1983. New Work for a Theory of Universals. *Australasian Journal of Philosophy*. 61(4): 343–377.

Lewis, David. 1986a. *Philosophical Papers*, vol. 2. Oxford: Oxford University Press.

Lewis, David. 1986b. *On the Plurality of Worlds*. Oxford: Blackwell.

Lewis, David. 1993. Many, but Almost One. *Ontology, Causality, and Mind*. K. Campbell, J. Bacon, and L. Reinhardt, eds. Cambridge: Cambridge University Press, 23–38.

Lewis, Peter. 2004. Life in Configuration Space. *British Journal for the Philosophy of Science*. 55(4): 713–729.

Lewis, Peter. 2013. Dimension and Illusion. In Ney and Albert 2013, 110–125

Lewis, Peter. 2016. *Quantum Ontology*. New York: Oxford University Press.

Loewer, Barry. 1996. Humean Supervenience. *Philosophical Topics*. 24: 101–127.

Maldacena, Juan and Leonard Susskind. 2013. Cool Horizons for Entangled Black Holes. *Fortschritte der Physik*. 61: 781–811.

Markosian, Ned. 2014. A Spatial Approach to Mereology. *Mereology and Location*. S. Kleinschmidt, ed. Oxford: Oxford University Press.

Maudlin, Tim. 1994. *Quantum Non-Locality and Relativity*. Oxford: Blackwell.

Maudlin, Tim. 2007a. Completeness, Supervenience, and Ontology. *Journal of Physics* A. 40: 3151–3171.

Maudlin, Tim. 2007b. Why Be Humean? In *The Metaphysics within Physics*. Oxford: Oxford University Press, 50–77.

Maudlin, Tim. 2013. The Nature of the Quantum State. *The Wave Function: Essays on the Metaphysics of Quantum Mechanics*. A. Ney and D.Z. Albert, eds. Oxford: Oxford University Press, 126–153.

Maudlin, Tim. 2019. *Philosophy of Physics: Quantum Theory*. Princeton, NJ: Princeton University Press.

McKenzie, Kerry. 2017. Ontic Structuralism Realism. *Philosophy Compass*. 12(4): e12399.

Mermin, N. David. 1981. Quantum Mysteries for Anyone. *Journal of Philosophy*. 78(7): 397–408.

Monton, Bradley. 2002. Wave Function Ontology. *Synthese*. 130(2): 265–277.

Monton, Bradley. 2006. Quantum Mechanics and 3N-Dimensional Space. *Philosophy of Science*. 73: 778–789.

Monton, Bradley. 2013. Against 3N-Dimensional Space. In Ney and Albert 2013.

Myrvold, Wayne. 2015. What Is a Wave Function? *Synthese*. 192(10): 3247–3274.

Newton, Isaac. 2007. Letter to Bentley. The Newton Project. http://www.newtonproject.ox.ac.uk/view/texts/normalized/THEM00254

Ney, Alyssa. 2010. Are There Fundamental Intrinsic Properties? In *New Waves in Metaphysics*. A. Hazlett, ed. London: Palgrave Macmillan, 219–239.

Ney, Alyssa. 2012. The Status of Our Ordinary Three Dimensions in a Quantum Universe. *Noûs*. 46(3): 525–560.

Ney, Alyssa. 2013. Introduction. *The Wave Function: Essays in the Metaphysics of Quantum Mechanics*. Oxford: Oxford University Press.

Ney, Alyssa. 2015. Fundamental Physical Ontologies and the Constraint of Empirical Coherence: A Defense of Wave Function Realism. *Synthese*. 192(10): 3105–3124.

Ney, Alyssa. 2016. Grounding in the Philosophy of Mind: A Defense. In *Scientific Composition and Metaphysical Ground*. K. Aizawa and C. Gillett, eds. London: Palgrave-Macmillan, 271–300.

Ney, Alyssa. 2017. Finding the World in the Wave Function: Some Strategies for Solving the Macro-object Problem. *Synthese*. doi:10.1007/s11229-017-1349-4.

Ney, Alyssa. 2019. The Politics of Fundamentality. In *What Is "Fundamental"?* A. Aguirre, B. Foster, and Z. Meerali, eds. Cham: Springer, 27–36.

Ney, Alyssa. 2020a. Separability, Locality, and Higher Dimensions in Quantum Mechanics. *Current Controversies in Philosophy of Science*. London: Routledge, 75–90.

Ney, Alyssa. 2020b. Grounding and the Philosophy of Mind. *The Routledge Handbook of Metaphysical Grounding*. London: Routledge, 460–471.

Ney, Alyssa and David Z Albert, eds. 2013. *The Wave Function: Essays in the Metaphysics of Quantum Mechanics*. New York: Oxford University Press.

Ney, Alyssa and Kathryn Phillips. 2013. Does an Adequate Physical Theory Demand a Primitive Ontology? *Philosophy of Science*. 80: 454–474.

Nielsen, Michael A., and Isaac L. Chuang. 2016. *Quantum Computation and Quantum Information*. Cambridge: Cambridge University Press.

Nolan, Daniel. 2005. *David Lewis*. Montreal: McGill-Queen's University Press.

Norsen, Travis. 2010. The Theory of (Exclusively) Local Beables. *Foundations of Physics*. 40(12): 1858–1884.

Norsen, Travis. 2015. Are There Really Two Different Bell's Theorems? *International Journal of Quantum Foundations*. 1: 65–84.

Norsen, Travis, Damiano Marian, and Xavier Oriols. 2015. Can the Wave Function in Configuration Space Be Represented by Single-Particle Wave Functions in Physical Space? *Synthese*. 192(10): 3125–3151.

North, Jill. 2013. The Structure of the Quantum World. In Ney and Albert 2013.

Pearl, Judea. 2000. *Causality*. Cambridge: Cambridge University Press.

Pearle, Philip. 1976. Reduction of the State Vector by a Nonlinear Schrödinger Equation. *Physical Review D*. 13: 1061–1084.

Popper, Karl. 1934/2002. *The Logic of Scientific Discovery*. London: Routledge.

Preskill, John. 2019. Notes from Quantum Field Theory (1986–87). http://www.theory.caltech.edu/~preskill/notes.html#qft.

Przibram, K., ed. 1934. *Letters on Wave Mechanics*. New York: Philosophical Library.

Putnam, Hilary. 1975. What Is Mathematical Truth? In *Mathematics, Matter, and Method*. Cambridge: Cambridge University Press, 60–78.

Rosen, Gideon. 2010. Metaphysical Dependence: Grounding and Reduction. In *Modality: Metaphysics, Logic, and Epistemology*. R. Hale and A. Hoffman, eds. Oxford: Oxford University Press, 109–136.

Ruetsche, Laura. 2011. *Interpreting Quantum Theories*. Oxford: Oxford University Press.

Schaffer, Jonathan. 2009. On What Grounds What. In *Metametaphysics*. D. Chalmers, D. Manley, and R. Wasserman, eds. Oxford: Oxford University Press, 247–282.

Schaffer, Jonathan. 2010. Monism: The Priority of the Whole. *Philosophical Review*. 119(1): 31–76.

Schaffer, Jonathan. 2017. The Ground between the Gaps. *Philosophers' Imprint*. 17(11): 1–26.

Schrödinger, Erwin. 1927. Collected Papers on Wave Mechanics. New York: Chelsea.

Schrödinger, Erwin. 1935. The Present Situation in Quantum Mechanics. In *Quantum Theory and Measurement*. J. Wheeler and W. Zurek, eds. Princeton: Princeton University Press, 152–167.

Sebens, Charles T. 2015. Quantum Mechanics as Classical Physics. *Philosophy of Science*. 82(2): 266–291.

Shankar, R. 2012. *Principles of Quantum Mechanics*, 2nd ed. Cham: Springer.

Shimony, Abner. 1990. An Exposition of Bell's Theorem. *Sixty-Two Years of Uncertainty*. A.I. Miller, ed. *NATO ASI Series*, vol. 226. Boston: Springer, 33–43.

Shrapnel, Sally. 2019. Discovering Quantum Causal Models. *British Journal for the Philosophy of Science*. 70(1): 1–25.

Sider, Theodore. 2003. Maximality and Microphysical Supervenience. *Philosophy and Phenomenological Research*. 66(1): 139–149.

Sider, Theodore. 2011. *Writing the Book of the World*. Oxford: Oxford University Press.

Smolin, Lee. 2006. The Case for Background Independence. In *The Structural Foundations of Quantum Gravity*. D. Rickles, S. French, and J. Saatsi, eds. Oxford: Oxford University Press, 196–239.

Smolin, Lee. 2007. *The Trouble with Physics*. New York: Houghton-Mifflin.

Stein, Howard. 1989. Yes, But . . . Some Skeptical Remarks on Realism and Antirealism. *Dialectica*. 43: 47–65.

Suárez, Mauricio. 2015. Bohmian Dispositions. *Synthese*. 192(10): 3203–3228.

Teller, Paul. 1986. Relational Holism and Quantum Mechanics. *British Journal for the Philosophy of Science*. 37(1): 71–81.

Teller, Paul. 1989. Relativity, Relational Holism, and the Bell Inequalities. In *Philosophical Consequences of Quantum Theory: Reflections on Bell's Theorem*. J.T. Cushing and E. McMullin, eds. Notre Dame, IN: University of Notre Dame Press, 208–223.

Teller, Paul. 1995. *An Interpretative Introduction to Quantum Field Theory*. Princeton, NJ: Princeton University Press.

Unger, Peter. 1980. The Problem of the Many. *Midwest Studies in Philosophy*. 5(1): 411–468.

Valentini, Antony. 2010. De Broglie-Bohm Pilot-Wave Theory: Many Worlds in Denial? In *Many Worlds? Everett, Quantum Theory, and Reality*. S. Saunders, J. Barrett, A. Kent, and D. Wallace, eds. Oxford: Oxford University Press, 476–509.

Van Fraassen, Bas. 1991. *Quantum Mechanics*. Oxford: Clarendon.

Van Inwagen, Peter. 1990. *Material Beings*. Ithaca, NY: Cornell University Press.

Varzi, Achille. 2016. Mereology. The Stanford Encyclopedia of Philosophy (Spring 2016 Edition), Edward N. Zalta (ed.). http://plato.stanford.edu/archives/spr2016/entries/mereology/.

Von Neumann, John. 1932/1996. *The Mathematical Foundations of Quantum Mechanics*. Princeton, NJ: Princeton University Press.

Wallace, David. 2012. *The Emergent Multiverse*. Oxford: Oxford University Press.

Wallace, David. 2020. Against Wave Function Realism. In *Current Controversies in Philosophy of Science*. S. Dasgupta and B. Weslake, eds. London: Routledge.

Wallace, David and Christopher Timpson. 2010. Quantum Mechanics on Spacetime I: Spacetime State Realism. *British Journal for the Philosophy of Science*. 61(4): 697–727.

Wigner, Eugene. 1961. Remarks on the Mind-Body Question. In *The Scientist Speculates*. I.J. Good, ed. New York: Basic.

Wilson, Jessica. 2013. A Determinable-Based Account of Metaphysical Indeterminacy. *Inquiry*. 56(4): 359–385.

Wilson, Jessica. 2014. No Work for a Theory of Grounding. *Inquiry*. 57: 1–45.

Wiseman, Howard. 2014. Two Bell's Theorems of John Bell. *Journal of Physics A*. 47: 424001.

Wittgenstein, Ludwig. 1953. *Philosophical Investigations*. London: Macmillan.

Worrall, John. 1989. Structural Realism: The Best of Both Worlds? *Dialectica*. 43 (1–2): 99–124.

Zeh, H.D. 1970. On the Interpretation of Measurement in Quantum Theory. *Foundations of Physics*. 1(1): 69–76.

Zurek, Wojciech. 1991. Decoherence and the Transition from Quantum to Classical. *Physics Today*. 44(10): 36–44.

Zurek, Wojciech. 2003. Decoherence, Einselection, and the Quantum Origins of the Classical. *Reviews of Modern Physics*. 75(3): 715–775.

Index

Abbott, Edwin, 112
Albert, David, ix, 15n16, 25n20, 27,
     34–35, 45–47, 52, 77, 89, 136,
     153–154, 192n7, 193, 199, 202,
     207, 210–219, 225, 231, 234n6,
     236–238, 245
Allori, Valia, 56, 59–60, 125–126,
     129n17, 134n1, 173, 182–184,
     187n5, 188–192
Aristotle, 65
Aspect, Alain, 101

Baez, John, 158
Barbour, Julian, 178
Barnes, Elizabeth, 242
Barrett, Jeffrey, 32n32, 175
basis-independence, 161–165
Becker, Adam, xn1, 17n17
Bell, John, 19, 21, 27, 28n27, 42, 45,
     60, 96–98, 101–105, 120, 143
Bell's theorem, 101–104
Belot, Gordon, 75, 192n8
Bickle, John, 228n2
Bohm, David, 26–27, 50
Bohmian mechanics
     Everett-in-denial objection
          to, 44–46
     and locality, 101n8, 110
     marvelous point
          interpretation, 46–47
     pilot wave interpretation,
          37, 44, 98
     and primitive ontology, 57
     as a solution to the measurement
          problem, 26–28

Bohr, Niels, 9, 15–17
Born, Max, 10–11, 14n15, 47
Born rule, 3–4, 6, 10, 18
Brown, Harvey, 44n43, 192n8
Byrne, Peter, 32n32

Calosi, Claudio, 66n5
Cao, ChunJun, 158
Carnap, Rudolf, xin2
Carroll, Sean, 25n21, 33n34, 158, 162
Cartwright, Nancy, 171–172
Chalmers, David, 90n5
Chen, Eddy Keming, 236
Clarke, Samuel, 126–127
collapse of the wave function,
     17–18, 28–29
complementarity, 9–10, 16–17, 99
configuration space, 37–38, 41,
     135–136, 150
     versus a reduced state space,
          235–236
Conroy, Christina, 33n33
Copenhagen interpretation, 17, 131

De Broglie, Louis, 26, 37, 98
the De Broglie-Bohm theory. *See*
     Bohmian mechanics
decoherence, 203–206, 220–221, 246
Dennett, Daniel, 222n9
density matrix, 74, 162–163
determinism, 13
DeWitt, Bryce, 33n33
Dirac delta function, 4
Dürr, Detlef, 25n20, 27, 56–58,
     136n3, 173, 181–184

Einstein, Albert, 14, 98, 113, 122–125
Einstein-Podolsky-Rosen (EPR)
    thought experiment, 50, 98–
    104, 114–119
empirical incoherence, 175
entanglement, 50, 83
ER=EPR conjecture, 121
Esfeld, Michael, 69, 72, 212
Everett, Hugh, 24, 30
Everettian quantum mechanics, 25
    and locality, 105
    and macroscopic objects, 245–246
    as a solution to the measurement
        problem, 30–32, 203–207
existence monism, 66
explanatory gaps, 229–231

Feynman, Richard, 131
Fine, Kit, 68n8, 227, 229, 239n10
Fock space, 142–143, 161
Forrest, Peter, 75–76
Fourier transform, 7–9
Fraser, James, 145n11
French, Steven, 69, 71–72, 235n8
functionalism, 210–219, 221–226,
    237–238

Ghirardi, Giancarlo, 28, 30n30, 60
Goldstein, Sheldon, 25n20, 27, 44,
    56–60, 173, 181–184, 192–193
Griffiths, David, 2n2, 8n10,
    19n19, 144n8
grounding, 226–231
GRW theory
    and locality, 105–110
    and macroscopic objects, 245
    and primitive ontology,
        59–60, 191
    as a solution to the measurement
        problem, 28–30, 201–202
guidance equation, 27

Hamiltonian, 12–14, 214–215
Hazlett, Allan, 196

Healey, Richard, 178
Heil, John, 239n10
Heisenberg, Werner, 17n18, 47
Heisenberg uncertainty
    principle, 7, 11
Henson, Joe, 117–119
hidden variables, 26–27, 31, 101. *See
    also* Bohmian mechanics
Hilbert space, 162–162
Hiley, Basil, 27n26
holism, 65–68, 247
Hooker, C.A., 228n2
Howard, Don, 66, 81–82, 85, 94,
    113–119
Hubert, Mario, 76
Huggett, Nick, 158
Humean supervenience, 82–83, 94–
    96, 127–129

intuitions, 129–132
Ismael, Jenann, 65, 111–112, 229

Kim, Jaegwon, 211
Kleinschmidt, Shieva, 241n11
Kochen-Specker theorem, 101

Ladyman, James, 69–71, 129, 190
Lam, Vincent, 212
Laudan, Larry, 69
Leibniz, Gottfried, 126
Lewis, David, 127–129, 193, 200, 252
Lewis, Peter, xn1, 14n15, 25n21, 35,
    90, 112, 120–121
local beables, 42, 167
    and evidence, 174–177, 179–180
    as fundamental, 209–210
    and primitive ontology, 181–195
locality, 96–98, 120–127, 143
    argument for wave function
        realism (*see* wave function
        realism: argument from
        locality)
    failure of locality in quantum
        mechanics, 98–104

and gravitational influence,
126–127
and relativity, 120–122
Loewer, Barry, ix, 82n2, 129, 193,
202, 207
Lorentz, Hendrik, 47
Luty, Markus, 141n4

Maldacena, Juan, 121
many worlds, 32–33. *See also*
Everettian quantum mechanics
Marian, Damiano, 75
Markosian, Ned, 240n11
Maudlin, Tim, 58–59, 102n9, 121,
128n16, 173–177, 194, 208–210
McKenzie, Kerry, 69
McQueen, Kelvin, 236
measurement problem,
14–15, 19–24
solutions to, 25–33, 61
mereology, 240–241
Mermin, David, 102n9, 131
metaphysical indeterminacy,
242–244
Michalakis, Spyridon, 158
Monton, Bradley, 64–65, 207–208
Monton's challenge, 207–210, 221–
224, 231–235
Myrvold, Wayne, 72, 90–95, 123–
124, 134, 137, 152, 157

narratability, 136, 153–157
Newton, Isaac, 126–127
Nolan, Daniel, 128
Norsen, Travis, 75, 87n3, 98n7
North, Jill, 49, 55, 77–79,
226–227, 229

ontic structural realism, 69–72, 85
ontology, 1, 33–34
Oriols, Xavier, 76

partial instantiation, 242–244
particle in a box, 2–3

Pearl, Judea, 114n13
Pearle, Philip, 28, 30n30
Penrose, Roger, 30n30
Phillips, Kathryn, 61, 181n3
Plato, 243
Podolsky, Boris, 98
Pooley, Oliver, 94
Popper, Karl, 178n2
Preskill, John, 143n6
Problem of the Many, 200, 242
primary ontology, 58–59
primitive ontology, 56–62, 78–79,
85–86, 181–184
argument for, 187–188
versus fundamental ontology,
183–184
See also local beables: and
primitive ontology
priority monism, 65–66, 240
Putnam, Hilary, 69

quantum field theories, 136–137,
140–149
quantum gravity, 158, 178, 192
Quine, W.V., 239n10

relational holism, 68, 85,
115–116
Rickles, Dean, 235n8
Rimini, Alberto, 28, 30n30
Romano, Davide, 76
Rosen, Gideon, 227, 229
Rosen, Nathan, 98
Ross, Don, 69, 129
Rovelli, Carlo, 178
Ruetsche, Laura, 144n9

Schaffer, Jonathan, 65–66, 68n8, 111,
190n6, 227, 229–230
Schrödinger, Erwin, ix, 11, 47–48,
50, 53, 84, 252
Schrödinger equation, 12–15
scientific realism, ix, 10, 69
Sebens, Charles, 26n22, 95n6

separability, 81–83, 122–140
    argument for wave function
        realism (*see* wave function
        realism: argument from
        separability)
    failure of separability in quantum
        mechanics, 84–85
    Howard's account of, 82
    and Humean supervenience,
        127–129
Shankar, Ramamurti, 3n3, 6n6,
    19n19, 144n8
Shimony, Abner, 97
Shrapnel, Sally, 121
Sider, Theodore, 59n3, 68n8, 200
Singh, Ashmeet, 162
Smolin, Lee, 158, 160
spacetime state realism, 72–74,
    78n12, 138–140, 162–163
    and separability, 85,
        139–140, 164
spontaneous collapse theories, 28–
    30, 31. *See also* GRW theory
state vector representations, 5, 161
Stein, Howard, 70–71
Stern-Gerlach apparatus, 20–21
structuralism, 69–72. *See also* ontic
    structural realism
    argument for, 69–70
    spatial, 212
Suárez, Mauricio, 64–65
superposition, 5
Susskind, Leonard, 121
symmetries, 222, 231–237, 239

Teller, Paul, 67–68, 70, 85, 114–116,
    137, 145n11
Timpson, Christopher, xin2, 72–
    73, 85, 134–136, 138–139,
    150–157, 162–164, 219–220,
    222, 225
Tumulka, Roderich, 56, 59, 173

Unger, Peter, 200, 242
universal wave function, 41

Vaidman, Lev, 88n4
Valentini, Antony, 44n43
van Fraassen, Bas, 26n23
van Inwagen, Peter, 200, 241n12
Varzi, Achille, 240–241
von Neumann, John, 17–21

Wallace, David, xn1, xin2, 25n20–21,
    44n43, 72–73, 85, 134–140,
    150–157, 159, 162–164,
    192n8, 199, 203n2, 219–222,
    225, 246n15
wave function realism
    argument from entanglement,
        49, 52–55
    argument from locality, 104–113
    argument from separability, 87–
        96, 138–140
    basic characterization of, 34
    the constitution objection to,
        166–169, 222
    and holism, 67
    the macro-object problem
        for, 198
    prima facie case for, 34–36
    relation to Bohmian
        mechanics, 43–45
    and relativistic quantum theories,
        134–159
wave function, 2–3
    as effective metaphysics, 157–158
    as a field, 45–46
    as a multi-field, 74–76, 85
    as nomological, 44, 86, 192–194
    as a property, 64–65, 85
    as a ray in Hilbert space, 162–165
    simple field interpretation of, 53
Weber, Tullio, 28
Wheeler-DeWitt equation, 192

Wigner, Eugene, 2, 22
Wigner's friend, 22–24, 32n31
Williams, J. Robert, 242
Wilson, Jessica, 111n12, 228–229, 242–244
Wiseman, Howard, 96–98, 104
Wittgenstein, Ludwig, 20

Worrall, John, 69–70
Wüthrich, Christian, 158

Zanghì, Nino, 25n20, 27, 44, 56–60, 173, 181–184, 192–193
Zeh, H. Dieter, 203, 205
Zurek, Wojciech, 203, 206